CELLULAR AND MOLECULAR BIOLOGY OF HUMAN OOGENESIS, OVULATION AND EARLY EMBRYOGENESIS

FUNDAMENTALS, BIOMEDICAL AND CLINICAL IMPLICATIONS IN RELATION TO INFANT DISORDERS

CELLULAR AND MOLECULAR BIOLOGY OF HUMAN OOGENESIS, OVULATION AND EARLY EMBRYOGENESIS

FUNDAMENTALS, BIOMEDICAL AND CLINICAL IMPLICATIONS IN RELATION TO INFANT DISORDERS

Sardul S. Guraya

Ph.D., D.Sc, FNA, FNASc, FNAAgSc,
former Head Deptt. of Zoology,
Dean faculty of Basic Sciences
& Humanity, PAU, Ludhiana.

NEW AGE

NEW AGE INTERNATIONAL (P) LIMITED, PUBLISHERS

(formerly Wiley Eastern Limited)

New Delhi • Bangalore • Chennai • Cochin • Guwahati • Hyderabad
Jalandhar • Kolkata • Lucknow • Mumbai • Ranchi

Visit us at www.newagepublishers.com

Published by New Age International (P) Ltd., Publishers
First Edition: 2007

ISBN : 81-224-1968-2

Rs.350.00

C-06-09-1039

Printed in India at Chaman Offset, Delhi
Typesetter: Kalyani Computers, New Delhi.

PUBLISHING FOR ONE WORLD
NEW AGE INTERNATIONAL (P) LIMITED, PUBLISHERS
(formerly Wiley Eastern Limited)
4835/24, Ansari Road, Daryaganj, New Delhi - 110002
Visit us at **www.newagepublishers.com**

PREFACE

The basic knowledge of ovarian development and maturation with special reference to formation of primordial oocyte, oocyte growth maturation, ovulation, fertilization, early development and biomedical and clinical implication of aging changes in oocytes, which are discussed in this book has number of applied aspects in immediate of long-term health prospects. Our knowledge of cellular and molecular biology of these aspects in the humans are of fundamental and applied interest to a wide variety of academic, scientific, toxicological, biomedical, biotechnological, clinical, disciplines for *ensuring* the normal early development and differentiation of human embryoes. Therefore, these aspects are discussed with an inter-disciplinary approach to produce knowledge that will help us in the better understanding of developmental disorders of children which are caused by developmental, maturational aging, genetic and environmental conditions such as chemical, xenobiotic factors, poor nutrition high ambient temperature, stress, improper management and *inadequate* health of developing child. The objective of this book is therefore, to project a much needed comprehensive, correlative, consolidated and critical account of recent advances in the morphological including ultrastructural, histochemical, immunological, physiological, endocrinological aspects of structural chemistry and functional significance of primordial oocyte, oocyte growth, maturation, ovulation, fertilization and early development in aging women, with reference to biomedical and clinical implication of aging changes in ovarian primordial oocyte. Actually application of modern biological, molecular and immunological techniques as discussed in this book have revealed the changes in the primordial oocyte with aging of women and their effect on subsequent oocyte growth, maturation, ovulation, fertilization and early development of embryoes. To the best of my information that basic scientist and *clinicitions* (Except my research papers, reviews etc.) have not given any attention to these aspects in humans. Although one general book on the aging of human ovary was published in U.S.A. by UNO. Any disturbances in the oocytes as a result of aging of women can cause changes in primordial oocyte and their effects on the subsequent oocyte growth and maturation, ovulation, fertilization and early development.

This book is organized in seven chapters dealing with the cellular, histochemical, ultrastructural, biochemical, molecular, immunological, genetical and endocrinological aspects of primordial oocytes, oocyte growth, oocyte maturation, ovulation, fertilization, early development and biomedical and clinical implication of aging changes in oocytes, besides presenting through update reviews, the concepts and theories of these various aspects and extensive bibliographics on these various aspects of the cellular and molecular biology of primordial oocytes and their growth, maturation, ovulation, fertilization and early development, besides the biomedical and clinical implication of aging changes in oocytes and up-to-date extensive bibliographics on various aspects of cellular and molecular biology of these various *subjects*, the future research needs related to each chapter are also clearly outlined, therefore the various chapters are up-to-date reviews with multi-disciplinary approach which will continue to serve as an important source of references for investigators especially biomedical and clinical scientist for years to come clearly, still there exists great gap in our knowledge of the cellular and molecular biology of formation of structure and function of various components of primordial oocytes, and their growth into the oocyte and their subsequent maturation, ovulation, fertilization and early development *in vivo* and *in vitro* conditions in regard to subsequent biomedical and clinical implications of aging changes in oocytes.

The current problems on these various aspects in humans were immersed from the fundamental ideas and discussed presently in this book, which is actually designed for those scientists who are involved in basic research on these aspects in the humans and other mammals. It is hoped that this book will certainly provide stimulation to reproductive, cellular, molecular, developmental biologists etc.

To fill in the various gap in our knowledge which are pointed out in this book for investigations in future.

The research work on the *various aspects of* Reproduction was carried out by the author in the Department of *Obstetrics and Gynaecology,* University of *Kansas Medical Centre, Kansas City U.S.A.* (as a post-doctoral fellow of the population council, New York, U.S.A.), Women's hospital, University of Hamburg, Hamburg, Germany as a visiting scientist and in the regional advance research centre in human reproduction established by the Indian Council of Medical Research at the Punjab Agricultural University, Ludhiana under the Directorship of the author.

Writing such a comprehensive book on a very important subject which is not published previously by any scientist (except myself in the form of great research papers, reviews etc.) by a single author in one volume is a most difficult task and thus I am very grateful to Dr. R.K. Sharma, Deptt. of Zoology, Kurukshetra University, Kurukshetra for revising the various chapters and getting it typed, as well as for his correspondence with the publishers.

Thanks are also due to authors and copyright holders for permission to republish *some of their illustrations.* I thanks my son Dr. Hermeet Singh for arranging and sending recent literature on this subject from U.S.A. I also owed a lot to my family, wife Surinder and my sons Gurmeet, Hermeet, daughter Rupa and son-in-law Ratesh, daughter-in-law Mona for providing encouragement and help during the completion of this *most difficult task*, while I was staying for some months with them in U.S.A.

Finally I would like to thank *New Age International Private Ltd. Publishers,* New Delhi, for providing excellent co-operation during publication of this book.

Ludhiana **SARDUL S. GURAYA**

CONTENTS

Introduction

The general basic functions of the mammalian and human ovary are two-fold, the production of ova and the secretion and release of hormones (Guraya, 1971, 1974, 1998a, b, 2000a; Filicori and Flamigni, 1996; Fauser et al., 1999; Yen et al., 1999; Tevelde et al., 2000; Bulletti et al., 2001). The results of recent studies on the comparative cellular and molecular biology of ovary in mammals in relation to production of ova and steroid hormone synthesis in vivo and in vitro have been reviewed and discussed by the present author (Guraya, 2000a) and other authors (Fauser et al., 1999; Yen et al., 1999; Tevelde et al., 2000; Sharma et al., 2000; Bulletti et al., 2001). But the past some years have witnessed a revival of great interest in the study of cellular and molecular biology of primordial follicles or oocytes, and normal development, growth, differentiation, structure, and physiology of oogenesis oocyte growth and maturation, ovulation, and fertilization of human ova *in vivo* and *in vitro* to develop better strategies to solve the problems of infertility and embryological disorders in human and sub-human primates, which have been subjected to diverse techniques of electron microscopy, histochemistry, immunology, autoradiography, physiology, molecular biology, biochemistry *in vivo* and *in vitro* systems (Fauser et al., 1999; Yen et al., 1999; Vats, 1999; Tevelde et al., 2000; Bulletti et al., 2001; Sharma, 2001; Sharma and Vats 2002). The purpose of this book is, therefore, to summarize and integrate the results obtained with these diverse techniques in order to obtain a deeper insight into the basic subcellular, molecular and functional aspects of human primordial follicles, oogenesis (oocyte growth and maturation), ovulation, fertilization and early embryogenesis. Special emphasis will be laid on developmental processes involved in these aspects in human at the cellular and molecular levels. An attempt will be made to lay special emphases on these aspects *in vivo* and *in vitro* in women of variable age groups for the better understanding of effects of aging about which our knowledge at the cellular and molecular level is still very meagre in spite of reviews on aging of reproductive organs, (Taketani and Kawagre, 1993; Fauser et al., 1999; Yen et al., 1999; Tevelde et al., 2000; Bulletti et al., 2001). These authors have laid major emphasis on the aging of ovary but not on various cellular and molecular aspects of primordial follicles and its subsequent effects on oogenesis, ovulation, fertilization and embryogenesis *in vivo* and *in vitro* etc., which are discussed in the present book. Aging is an age-related decrement of functions in cells, tissues, organs or the whole body, therefore the time is just right to review the effects of aging on primordial follicles and its effects on the subsequent oocyte growth and maturation, ovulation, fertilization and early embryogenesis especially in humans because one third of women trying to have baby after the age of 35 will have problems with fertility and two thirds will not be able to get pregnant spontaneously once they past 40, according to the American Society of Reproductive Medicine. This is due to the fact that as early as 15 years before menopause, a woman egg production starts to decline. Actually the quality of eggs also becomes poor and more of the eggs she does produce contain chromosome problems. That make infertility, miscarriage and birth defects more. Actually, very little or no attention has been paid to the effects of aging of women on the cellular and molecular aspects of her primordial oocytes. If the data is not available on these aspects under discussion in the human then the observations available on these aspects in sub-human primates and other mammals will be taken into consideration to fill the gaps as relatively cellular and molecular aspects of aging of primordial follicles on oocyte growth and maturation, ovulation, fertilization and embryogenesis are not studied in human (Fauser et al., 1999; Yen et al., 1999; Tevelde et al., 2000; Bulletti et al., 2001). Such an integrated knowledge of human oogenesis, ovulation, and fertilization, and early embryogenesis, which has not been made available previously, is essential for a better understanding of the consequences of the influences not only of aging but also of chemicals (including drugs, free radicals, alcohol, etc), metals, radiation, heat, stress, and other chemical and physical agents to which the ova of human and subhuman primates and other mammals are being increasingly exposed at various stages of their growth and maturation *in vivo* and *in vitro* conditions in order to develop the possibilities and strategies of solving the problem of human fertility and sterility in the female and of developmental abnormalities of the foetus (Fauser et al., 1999; Yen et al., 1999; Tevelde et al., 2000; Bulletti et al., 2001). The campaign sponsored by The American Society for Reproductive Medicine actually addresses several risk factors for infertility. The detailed account of the cellular and molecular aspects of primordial oocyte growth, and maturation of oocyte, fertilization and embryo development after fertilization in human and sub-human primates is required to provide rational interpretation of alterations caused by aging of primordial oocytes and ovulated ova and prolonged action, even in weak doses of different types of chemicals, drugs, stress and radiation (Himms-Hagen, 1999) and smoking (Sharma, 2001; Sharma and Vats 2002) on human oocytes. Previous reviews deal mostly in isolation with the cellular and molecular aspects of primordial oocytes, oocyte growth and maturation, ovulation, fertilization and early, embryogenesis in various other mammalian species (Thibault et al., 1993; Folicori and Flamigni, 1996, Guraya, 2000a; Vats,1999; Sharma, 2001; Sharma and Vats 2002) paying very little or no attention to integrate the results of various modern techniques of cellular and molecular biology, which are being obtained for human *in vivo* and *in vitro* conditions (Fauser et al., 1999; Yen et al., 1999; Tevelde et al., 2000; Bulletti et al., 2001). The gaps in our knowledge about the effects of cryopreservations and culturing of follicles or oocytes, fertilized eggs, and early embryos *in vitro*, which are being carried out in human these days (Wolf and Zeilinski-Wooten, 2001), are required to be studied at the cellular and molecular levels as will be pointed out here.

Morphological, histochemical, biochemical and physiological especially cellular and molecular changes in the membrana granulosa, follicular fluid and the thecae of follicles in humans and other mammals, which are associated with the pre-ovaulatory swelling and ovulation have been discussed in detail in previous review articles and books published over the past some years (see references in the Thibault et al., 1993; Folicori and Flamigni, 1996; Yen et al., 1999; Fauser et al., 1999; Guraya, 2000a; Sharma et al., 2000) and thus will not be discussed. Here a major emphasis will be laid on the cellular and molecular aspects of ooplasmic and nuclear components of primordial oocytes, oocyte growth, and maturation, and ovulated ova, fertilization and early embryogenesis in women of different ages under *in vivo* and *in vitro* conditions. After integrating the results of various studies on cellular and molecular biology of primordial oocytes, oogenesis (oocyte growth, maturation and ovulation), fertilization and early embryos in relation to age of women, gaps in our knowledge about these aspects will be pointed in each chapter for further investigations to solve the problems of infant disorders in aged women.

Although the techniques of cryotechnology and culturing are being used to study various aspects of mammalian and human reproduction and development (Wolf and Zeilinski-Wooten, 2001) but these are not yet perfected, the major question now is how to use the tissue most effectively after thawing (Oktay et al., 1998; Fauser et al., 1999; Menezo et al., 2000; Menezo and Ben-Khalifa, 1995). For the present ovarian tissue cryopreservations and culturing are especially at the experimental stage, but these hold the promise of valuable applications in human oocyte growth, and maturation, ovulation, *in vitro* fertilization early embryogenesis etc. (Abir et al., 1999; Gook et al., 1999; Rutherford and Gosden, 1999; Fauser et al., 1999; Yen et al., 1999; Tevelde et al., 2000; Bulletti et al., 2001). Recently, cryosurvival of the zygotes and cleaved embryos has been found unsatisfactory and thus cryopreservations of *in vitro* matured embryos may not be an optimal procedure (Menezo et al., 2000; Suikkari et al., 2000; Wolf and Zeilinski-Wooten, 2001).

1

Primordial Oocytes

Taking into consideration the normal reproductive life up to 40 years or more of women, some primordial oocytes continue to remain quiescent in the ovarian cortex for four decades or more (Foote, 1975). Meanwhile, they are expected to undergo aging changes in terms of somatic cell life, as aging is an accepted property of multicellular organisms as well as of their gametes (Blandau 1975; Sauer, 1984). It was pointed out that fertilization of aged or devitalized egg may lead to abnormal development that may express itself in death and spontaneous abortion or resorption of the embryo, or more tragically in the birth of a child with developmental abnormalities, disorders or with subtle deficiencies, involving the full range of mental and physiological retardations (Vats, 1999). This suggestion was made on the basis of aging of fully differentiated oocytes or eggs during the preovulatory period or after ovulation if not fertilized in time. Very little or no attention has been paid to the aging changes of primordial follicles (or oocytes) and their subsequent biomedical and clinical implications (Blandau, 1975; Motta *et al.*, 1995; Filicori and Flamigni, 1996; Guraya, 1999a) which will be discussed in this book.

Primordial oocytes in women may remain quiescent for more than 40 years or more before ovulation. Associated with this, is marked increase in frequency of Down's syndrome or mongolism (Edwards, 1970a). The course of this genetic defect was believed either impairment with age of the placement of chromosomes on the spindle during the first meiotic division or impairment of the process of ovulation leading to oocytes that deteriorate just prior to fertilization. A third hypothesis was suggested that time at which oocytes are formed affects the time of life they will be ovulated and those formed last carry a greater risk of having chromosomal abnormalities (Edward, 1970b). Thus, oocytes ovulated in older women would come from oocytes formed abnormally. Very little attention has been paid previously to the aging changes of primordial follicles (or oocytes), which will be described here as these may lead to their partial deterioration or to loss of vitality as also speculated by Hertig and Adams (1967) who, of course, obtained primordial follicles from women 28-37 years of age for their electron microscopic studies but did not make any correlation of structural changes in the ooplasmic components to the age of women. Guraya (1967a, 1970b, 1974, 1999a, b) was the first worker who paid attention to this aspect. Therefore, it is felt essential to give a detailed critical review of histochemical and ultrastructural aspects of these quiescent primordial follicles in the human ovary with focus on their aging changes.

Then the possible biomedical and clinical implications of these aging changes in relation to human disorders in development and physiology will be discussed in chapter 7 as no attempt has been made previously in this regard (Filicori and Flamigni, 1996). Based on this discussion a hypothesis will be put forth to stimulate further investigations with modern immunocytochemical and molecular probes to approve or disapprove it. This is the major objective of this chapter as studies have not been carried out previously in this regard and are thus very timely. For the better appreciation or understanding of formation, structure and aging changes of quiescent primordial follicles, it will be desirable first to give a brief account of the origin, migration, proliferation and differentiation of female germ cells (primordial germ cells and oogonia) and then to discuss the differentiation of oogonia into primordial oocytes during the development of the human ovary. Germ cell loss or reduction forming the characteristic feature of developing, maturing and cycling ovaries of human will also be described as major portion of germ cells is lost mainly by degenerative process (or atresia).

1.1 ORIGIN AND MIGRATION OF PRIMORDIAL GERM CELLS, AND FORMATION OF PRIMORDIAL FOLLICLES (OOCYTES)

With the development of human ovaries during the foetal life, about 6-7 million primordial follicles (oocytes) are formed from the oogonia, which in turn derive from the primordial germ cells (PGCs) of the early embryo (Baker and O, 1976; Guraya, 1998a; Fauser *et al*,1999). The PGCs first originate from the yolk sac stalk epithelium before gonadal differentiation and even before the development of the gonadal anlage. At this stage PGCs appear larger, clear and stain intensely for alkaline phosphatase (Motta *et al*., 1995, 1997a, b; Guraya, 1999a). They are generally rounded and possess a diameter ranging between 15 and 20 μm. Their nucleus is in an eccentric position and contains fine granular chromatin, uniformly dispersed within the nucleoplasm showing one or two large nucleoli. The cytoplasm of PGC is relatively poor in organelles as oval or rounded-shaped mitochondria having tubulo-vesicular cristae, a single Golgi complex, membranes of the rough endoplasmic reticulum (RER) often close to the perinuclear spaces, free ribosomes, polysomes and vesicles can be seen, together with a few micofilaments; centrioles and microtubules are also present. In addition, glycogen particles and lipid droplets are of common occurrence in human PGC cytoplasm especially at the starting of its migration to the genital ridges, which are believed to form an energy reserve to be used during this migration (Guraya, 1998a). Focal areas of close contact between PGCs and the neighbouring somatic cells can be seen as evidenced by the presence of desmosomes, intermediate junctions and tight junctions in these areas, suggesting metabolic coupling between them (Makabe *et al*., 1989, 1991).

PGCs, actively proliferating, migrate from the yolk sac epithelium to the gonadal anlage through the hind gut (Guraya, 1998a,1999a; Fauser *et al,* 1999). Around the fourth week post fertilization (p.f.), in human, numerous PGCs are observed in the hind gut epithelium and this is the result of their passive translocation (Matta *et al*., 1995,1997a,b). Active migration of PGCs occurs through the dorsal mesentery during the fifth week of embryo development by developing protrusions and pseudopodia of their plasma membrane. Meanwhile their former rounded shape, with well-defined contours, develops more irregular features and become spindle-shaped cells, having a long axis reaching 30 mm. The cytoplasmic features are also altered as the nuclear envelope becomes somewhat irregular, membranes of RER increase in number and delimit large intramembranous cisternae, and many microtubules and microfilaments appear especially in the cortical regions and within the cell protrusions (Makabe *et al*., 1991; Motta *et al*., 1995, 1997a, b). Active migration of PGCs is clearly the result of their own amoeboid movements. Fibronectin and other components of the extracellular matrix are believed to contribute to the migration of PGCs as the free surface of migrating PGCs and especially their pseudopodia are covered by a delicate fibrillar coat which corresponds to a glycocalyix; the latter appears to be associated with the binding sites of specific macromolecular components of the extracellular matrix, including fibronectin. This special coat of PGCs plays an important role in their adhesion to the substrate as well as in their recognition of the migratory route towards the gonadal anlage (Pereda and Motta, 1991; Guraya, 1998a, 1999a). Still some of them leave the right way during migration, thus reaching ectopic positions where most of them degenerate but some of the ectopic cells may differentiate and may also be related to the occurrence of some tumours in the adult (Guraya, 1998a, 1999a; Fauser *et al,* 1999).

At the end of the fifth week of development, or during the sixth one, PGCs reach the gonadal anlage, colonizing the most superficial zones of the developing ovary (Makabe *et al*., 1991; Guraya, 1998a; Fauser *et al.,* 1999). Gonadal PGCs still retain their motile capability wandering through the ovarian tissue as evidenced by the presence of some amoeboid processes. Within the cytoplasm, glycogen particles and lipids are very few; small mitochondria, polyribosomes, endoplasmic retiouluns (ER) and Golgi elements are in turn well represented. The alkaline phosphatase activity is still present. After reaching the gonadal tissue, the PGCs rapidly increase in number by mitosis and meanwhile there occurs a hyperplasia of the surrounding somatic components.

Actually the genital ridges are formed by different types of somatic cells: a proliferating coelomic epithelium covering the developing gonad and an underlying compartment having mesenchymal cells, blood vessels and mesonephric cells originating, via the rete system, from the mesonephric glomeruli and tubules (Guraya, 1998a). But which of these components of the developing gonad plays the major role in the formation of the ovarian blastema is still being debated (Wartenberg, 1989; Guraya, 1998a, 1999a). With the increase in the size of the developing ovary, there occurs an intermixing of the local somatic cells and the foreign germ cells, resulting in the formation of cellular cord-like structures in the ovarian cortex. The fragmentation of these germ-somatic cell aggregates will lead to the formation of primordial follicles which first develop in the inner regions of the ovarian cortex.

During further foetal life of the female the PGCs undergo repeated mitotic divisions to form oogonia. Germ cell differentiation starts in the cortical regions of the developing ovary (in humans it occurs around the ninth week p.f.) with the formation of oogonia from proliferating PGCs (Wartenberg, 1989; Motta *et al*., 1995, 1997a, b; Guraya, 1998a, 1999a; Fauser *et al.,* 1999). As compared to PGCs the oogonia show more regular and smooth outline and possess a large and spherical nucleus which is placed at the centre of the cell and contains little chromatin and one to three distinct reticular nucleoli. Their cytoplasm is reduced to small rim, containing a reduced number of organelles such as Golgi membranes, free ribosomes, round and oval mitochondria usually placed in a perinuclear arrangement and scanty elements of ER. Lipid inclusions and glycogen present in the PGCs appear to diffuse suggesting their utilization for metabolic and developmental processes in the PGCs. But alkaline phosphatase activity continues to be present. The most conspicuous features of oogonia are irregular bridges interconnecting a clonial association of oogonia suggesting communication between them. These sort of germ-cell syncytia are believed to derive from the incomplete divisions of the cell body during the subsequent rapid mitotic divisions (Guraya, 1998a, 1999a). Various organelles can be seen in the narrow rim of cytoplasm of the bridge. By the intercellular communication, intercellular bridges appear to coordinate the differentiation and/or degenerative processes that affect the germ-cell line inside each nest (Guraya, 1998a). The mitotic divisions of oogonia are synchronized and clusters of dividing oogonia showing identical chromosomal configurations form a common feature. Desmosome-like structures and small gap junctions are seen on oogonia and adjacent somatic cells (Motta *et al.*, 1995, 1997a,b; Guraya, 1998a,1999a). The intermediate stage of germ cell differentiation differing in some respects from oogonia and PGCs is also observed as the intermediate type of germ cells are less dense than PGCs and do not show the very light appearance of proliferating oogonia (Wartenberg, 1989). After losing the structural attributes of PGCs (glycogen, cytoplasmic processes), the intermediate type of germ cell still does not show all the characteristics of oogonia, such as oogonial clusters, synchronized mitosis and intercellular bridges connecting identical oogonia during their interphase.

After numerous mitotic divisions proliferating oogonia located in the inner cortex of the ovary begin to differentiate into oocytes (Wartenberg, 1989; Makabe *et al.*, 1991; Motta *et al.*, 1995, 1997a,b; Guraya, 1998a,1999a; Fauser *et al.*, 1999). Then meiosis starts usually in the innermost regions of the ovarian cortex; in humans it begins during the 12th-13th week p.f. (Motta *et al.*, 1995, 1997a,b; Fauser *et al.*, 1999). As described for oogonia, the nests of oocytes joined by intercellular bridges are not uncommon. Meiosis begins at this developmental stage.

The oogonia pass through the early stages of meiotic prophase (leptotene to pachytene) to form the primordial oocytes (Guraya, 1998a; Fauser *et al.*, 1999). Regulatory factors involved in the initiation of meiosis in oogonia as well as in the meiotic metaphase arrest and activation are still to be defined more precisely at the molecular level (Byskov *et al.*, 1995, 1999b; Whitaker, 1996; Sagata, 1996; Downs, 1996; Guraya, 1998a,1999a; Fauser *et al*, 1999). Adashi (1990) has suggested that the initiation as well as the arrest of meiosis in oocytes may be under in situ paracrine and autocrine control of growth factors. The control of meiotic arrest can be usefully formulated in terms of the interaction between the cell signalling mechanisms and the protein machinery that controls the cell cycle proteins (Whitaker, 1996). Germ cell size increases as oocyte development progresses towards the diplotene stage. Important changes occur in the oocyte nucleus due to triggering of the meiotic process (Fauser *et al.*, 1999). In the cytoplasm, mitochondria become more numerous and lie along the outer surface of the nuclear membrane. Golgi complex is also present near the nucleus encircling the centriole. It is believed that this nuclear polarization of organelles required for oocyte metabolism at this early stage depends upon microtubular activity (Motta *et al.*, 1997). Membrane-bound dense bodies can be seen in the cytoplasm (Motta *et al.*, 1995, 1997a,b). Alkaline phosphatase disappears after germ cells enter meiosis.

The differentiation of oogonia into oocytes is closely accompanied by their association with pregranulosa or follicular cells which get separated from the somatic cells of the ovarian blastema (Makabe *et al.*, 1991; Motta *et al.*, 1995, 1997a, b; Guraya, 1998a,1999a). It is still a matter of debate whether these cells originate from the neighbouring mesonephros or solely from ingrowths of coelomic epithelium (Guraya, 1998a). Apart from their actual origin, these somatic cells proliferate and intermingle with the germ cells to form pre-follicular cells, which continue to surround the germ cells that are near to being oocytes; even when intercellular bridges are eliminated and fragmentation of the nests occurs, the somatic cells play an active role in this fragmentation (Guraya, 1998a,1999a). With these events a true folliculogenesis starts. This occurs in human foetuses from 17-20 weeks of gestation until term. The first primordial follicles (or oocytes) develop at the inner border of the cortex or between the medullary cords during the 18th week of gestation. Finally most of the former compact cortex of the ovary is replaced by primordial follicles during the fifth month. In these follicles, desmosomes and small gap junctions can be seen among the germ and surrounding follicular cells (Guraya, 1998a, 1999a). In addition, a basal lamina now separates these cells (that will differentiate into pregranulosa cells from the ovarian blastema) thus separating the germ-follicular cell complexes (Makabe *et al*., 1991; Guraya, 1998a, 1999a). The somatic cells of mesenchymal origin will further associate to constitute the future thecal folliculi. Oogenesis continues as long as the remnant persists, but generally it is terminated at the end of the seventh month in the human. Besides the increasing number of resting primordial follicles, some growing follicles also appear during the foetal ovary. Finally numerous primordial oocytes enveloped by some follicular cells are formed and occupy the ovarian cortex at term and during the prepubertal and pubertal periods (Fig. 1) (Makabe *et al*., 1991; Motta *et al*., 1995, 1997a, b; Guraya, 1998a,1999a).

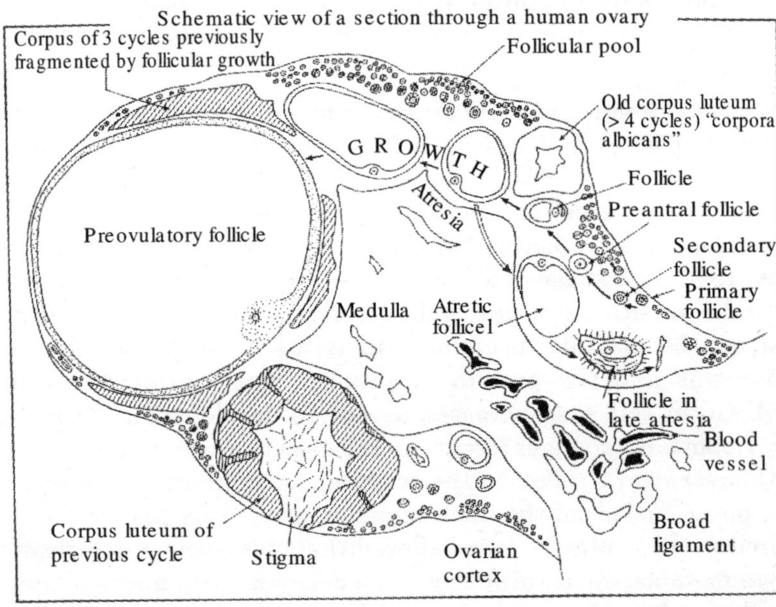

Fig. 1 Schematic view of a seation through a human overy

1.2 GERM CELL LOSS OR REDUCTION

The number of PGCs that left the yolk sac being few hundred only, rapidly increases by mitosis during their migration to the gonadal anlage and later, after their settlement in the human ovary, oogonia, which derive from the PGCs, not only differentiate into oocytes, but also undergo a remarkable mitotic proliferation so that there are about 600,000 by eighth weeks of gestation and 7,000,000 by the fifth month of gestation (Baker and O, 1976). Most of these oogonia degenerate, but some live, enter meiosis, and become oocytes, which get associated with follicular cells to form primordial follicles. Thus, only a very small number of germ cells will ovulate during the reproductive life, a major portion of them are lost through a degenerative process (Fig. 1) (Rabinovici and Jaffe, 1990; Byskov and Hoyer, 1994; Motta *et al.*, 1995, 1997a,b; Guraya, 1998a, 2000a; Fauser *et al.*, 1999). However, a small portion of them is also lost by elimination through the surface epithelium (Motta and Makabe, 1986; Motta *et al.*, 1995, 1997a, b; Guraya, 1998a, 1999a). A very impressive germ cell loss occurs during the prenatal development of the ovary (Guraya, 1998a, 2000a). Actually throughout the reproductive period, the human ovary is a dynamic organ that may change in size, shape, weight, chemistry, physiology at each stage of the menstrual cycle (Guraya, 1971, 1974; Netter, 1993; Suganuma *et al.*, 1993) and molecular biology (Fauser *et al.,* 1999). There is seen a gradual disappearance of primordial ova and follicles and a generalized necrosis of 7 million oocytes formed by midgestation, only 2 million are present at birth and only about 300,000 at menarche (Netter, 1993; Sugunuma *et al.*, 1993), which undergo aging changes as already poined out. The menopause is undoubtedly due to failing ovarian function because of the complete exhaustion of the pool of primary follicles and associated decreased steroidogenesis: Gosden et al. (1996), have discussed the biological bases of premature ovarian failure. The ovary is endowed at birth with a fixed number of primordial follicles, which steadily dwindles throughout life as a result of atresia and recruitment towards ovulation (Fig. 1). A causal relationship between follicle depletion and menopause clearly exists, and there is a gradual acceleration of follicle wastage in the human ovary beginning more than a decade before the end of menstrual life. A mathematical model has provided confirmatory evidence of this relationship, and indicates that menopause is triggered by a threshold number of follicles which varies stochastically with a mean of 1100.

Degenerative process is very extensive in oogonia undergoing mitosis and oocytes in meiotic prophase in human (Makabe *et al.*, 1989, 1991; Robinovici and Jaffe, 1990; Guraya, 1998a). Degenerative changes are of common occurrence in the stages preceding primordial follicle formation in the human (Rabinovici and Jaffe, 1990). A relatively high incidence of synaptic errors is observed in human oocytes during the pachytene stage when compared to other species (Robinovici and Jaffe, 1990; Guraya, 1998a). The high rate of germ cell atresia in the foetal ovary of human is partly attributed to these meiotic pairing anomalies. The number of degenerating oogonia in interphase and mitosis is insignificant during human oogenesis. But there occurs an increase in atresia after the end of first trimester when meiotic changes (prophase I) are well under way (Rabinovici and Jaffe, 1990; Guraya, 1998a). After their incorporation into primordial follicles, a few oocytes undergo degeneration (Baker and Franchi, 1967a). It is not known whether meiotic pairing anomalies are causal or consequential to degeneration. The most common degenerative signs observed are swollen nuclei, changes in the nuclear profile, condensation of chromosomes, mitochondrial damage, vacuolization, and dilation of ER (Motta *et al.*, 1995, 1997a, b); an increase in the volume fraction of the smooth ER and vacuoles in the oocytes of atretic primordial follicles compared with intact oocytes is also observed.

The regulatory factors involved in causing extensive degeneration of germ cells during the development of the ovary are required to be determined more precisely at the molecular level (Motta *et al.*, 1997b; Guraya, 1998a; Fauser *et al*, 1999). The degenerative process may be the result of genetic errors occurring during crossing-over involved in the meiotic process, as well as of metabolic and/or vascular disturbances. It may also extend to affect the surrounding follicular cells, resulting in true follicular atresia. The molecular basis of ovarian cell death during germ cell attrition (follicular atresia) is of great current interest (Amsterdam *et al.*, 1996; Eisenhauer *et al.*, 1996). Since a few primordial oocytes in the human ovary undergo degeneration, the tremendous loss of germ cells is the result of degeneration of primordial germ cells and differentiating oogonia, leading in humans to a critical reduction in number of germ cells from seven millions during the fifth month of intrauterine life to about 500,000 primordial follicles at birth and this is all that a female will have for the rest of her life (Block, 1952; Fauser *et al.*, 1999).

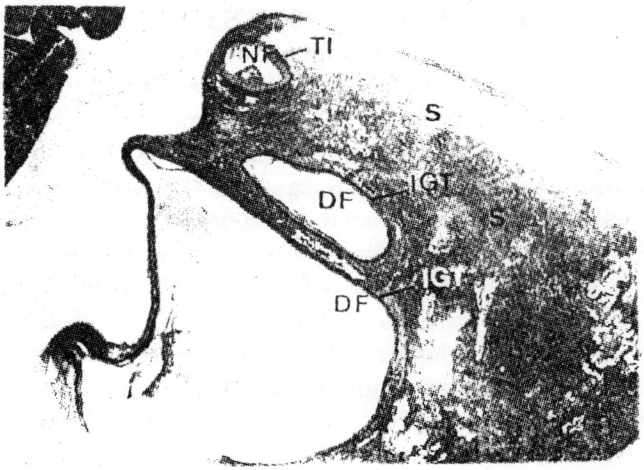

Fig. 2

After birth, the number of primordial follicles continues to decrease during the ovarian maturation and reproductive life of human owing to the repeated initiation of growth in some of them. Factors involved in the initiation of follicular growth are still required to be investigated at the molecular level (Gougeon, 1996a, b; Guraya, 1998a, 2000a; Fauser *et al.*, 1999). Various types of follicles have been classified. Follicular types in the human ovary are classified as primordial follicles (with only flattened follicular or pregranulosa cells), follicles with a mixture of flattened and cuboidal granulosa cells, primary follicles (one layer of cuboidal granulosa cells), early growing follicles or secondary follicles (without antrum and without theca interna), preantral follicles (without antrum and differentiated theca enterna) antral follicles or tertiary follicles and preovulatory follicles (Fig. 1) (Guraya, 1985, 1998a, 2000a; Hirshfield, 1991; Motta *et al.*, 1994; Hsueh *et al.*, 1994; Gougeon, 1994, 1996a,b; Fauser *et al.*, 1999). These follicles terminate with maturation and ovulation (Fig. 1) (Guraya, 1968, 1971, 1974; 1985, 2000a; Makabe *et al.*, 1989, 1991; Motta *et al.*, 1995) or with atresia (Fig. 1) (Guraya, 1966, 1867b, 1971, 1985, 1998, 2000a; Hsueh *et al.*, 1994). Sankova *et al.* (1985) observed statistically significant differences in the volume density of mitochondria, smooth ER and Golgi apparatus between the oocytes of intact primordial and primary follicles of human ovaries. Follicular atresia forms the predominant feature of human ovary during the menstrual cycle, pregnancy and lactation (Figs. 1 and 2) (Guraya, 1966, 1967b, 1971, 1972, 2000a) as well as during the early postmenopausal years (45 to 52 years of age) (Guraya, 1976a). As a result of cyclic growth of primordial follicles, only a few of them are left behind in the ovarian cortex of early postmenopausal women (Guraya, 1976a, b).

1.3 STRUCTURAL AND HISTOCHEMICAL ASPECTS OF PRIMORDIAL FOLLICLES

The correlation between the findings of electron microscopic and histochemical studies on the human primordial follicles in young and old women has shown that they are round structures consisting of quiescent oocyte with spherical nucleus and flattened follicular cells separated from the surrounding vascularized and innervated ovarian stroma by a delicate basal lamina (Figs. 3A and 3B) (Guraya, 1967a, 1970, 1971, 1985, 1999a, 2000a; Makabe *et al.*, 1989, 1991; Motta *et al.*, 1994, 1995, 1997a, b).

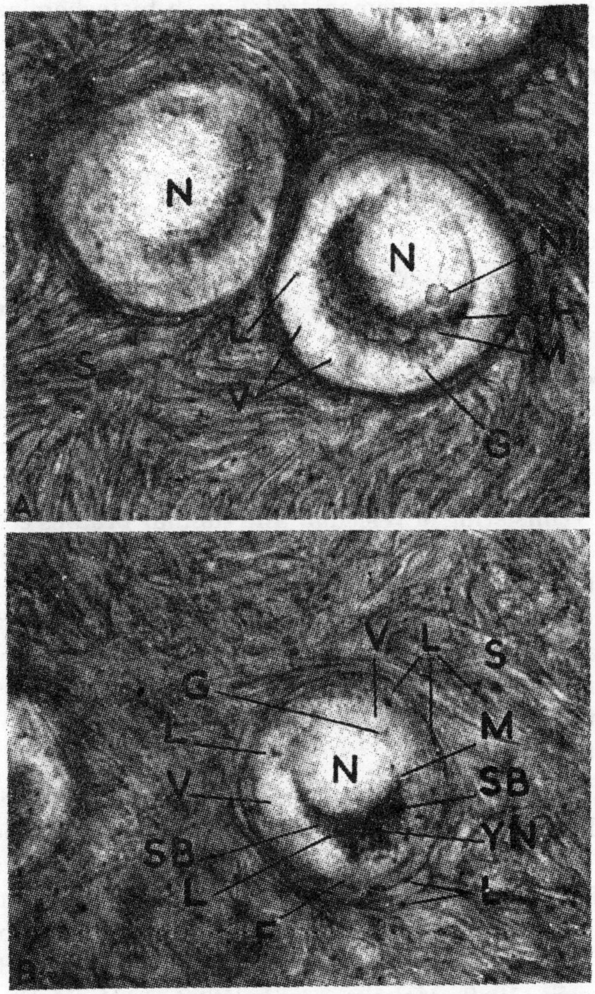

Fig. 3

1.3.1 Ooplasmic Components

The oocyte having a spherical or ovoidal shape shows a complex paranuclear structure called as the Balbiani's vitelline body or yolk nucleus complex (Hertig and Adams, 1967; Hertig, 1968; Guraya, 1961. 1967a, 1970, 1974, 1985. 1999a, 2000a). It consists of yolk nucleus or cytocentrum or archoplasm, granular basophilic substance containing some elements of lipoprotein, mitochondria, spherical bodies, Golgi complexes, annulate lamellae and lipid bodies (Figs. 3A, 3B and 4A, 4B, 4C). Stankova *et al.*, (1985) while making morphometric study of human primordial follicle, have not distinguished various components of Balbiani's vitelline body, only smooth ER, mitochondria and Golgi apparatus are taken into consideration. The cytocentrum forming a juxtanuclear mass consists of RNA, proteins and some lipoproteins. By electron microscopy, it shows a central aggregate of amorphous electron opaque deposists, which appear to become periodically aligned on fine fibrils to constitute the long coarse fibres at the periphery of the cytocentrum (Figs. 5A and 5B) (Stegner, 1967; Hertig and Adams, 1967; Hertig, 1968). The cytocentrum also shows vesicles of variable size (Figs. 5A, 5B), which consist of lipoproteins. According to Guraya (1970, 1999a, 2000a), the RNA and proteins demonstrated with histochemical techniques form the ultrastructural electron-opaque deposits, while the lipoproteins may be due to fibrils and vesicular elements seen in the cytocentrum (Fig. 5A, 5B) which was not distinguished by other workers (Baca and Zamboni, 1967; Stankova *et al.*, 1989, 1991; Motta *et al.*, 1994, 1995) who appear to have clumped it with the Golgi complex. The well-differentiated cytocentrum or yolk nucleus is also described as the "archoplasm" or idiosome (Guraya, 1974, 1999a) has not been reported in the oogonia and oocytes of foetal prematurely born, neonatal and early postnatal specimens (Fig. 5 D). (Stegner and Wartenberg, 1963; Lanzavecchia and Mangioni, 1964; Baker and Franchi, 1967a; Stegner, 1967; Makabe *et al.*, 1989, 1991; Motta *et al.*, 1995, 1997a,b). Based on the differences, Guraya (1967a, 1970, 1999a) suggested that the cytocentrum or yolk nucleus differentiates late in human primordial oocytes by the aggregation of ribonucleoproteins, fibrils and lipoprotein elements of ER. Its development is clearly related to the development and activity of decondensed chromosomes in the nucleus of primordial oocytes (Baker and Franchi, 1967a, b; Guraya, 1974, 1999a) as will also be described later.

The Golgi complexes consisting of vesicular and tubular profiles composed of lipoproteins are present in close association with the periphery of cytocentrum (Figs. 5A and 5B) (Hertig and Adams, 1967; Hertig, 1968; Guraya, 1999a, 2000a). In some primordial oocytes they appear to form a fenestrated shell around the cytocentrum. The Golgi complexes are also interspersed among the mitochondria and in addition are frequently associated with ER near the oolemma of primordial oocyte. The other workers using electron microscopy appear to have lumped the cytocentrum and its associated Golgi elements under the term Golgi complex as already stated.

The granular basophilic substance is distributed mainly around the yolk nucleus or cytocentrum (Figs. 4A, 4B and 4C) and consists of proteins, RNA and some lipoprotein filaments (Guraya, 1967a, 1970, 1974, 1999a, 2000a). Ultrastructurally, it consists of ribosomes and some elements of granular and smooth ER taking the form of vesicles and cisternae (Figs. 5A, 5B and 6A, 6B), which in the case of granular ER are sporadically and irregularly studded with ribosomes (Hertig and Adams, 1867; Stegner, 1967; Hertig, 1968; Baker and Franchi, 1867a; Dvorak and Tesarik, 1980; Makabe *et al.*, 1989, 1991; Motta *et al.*, 1995, 1997a) polysomes are inconspicuous. Microtubules lie most frequently around the nuclear envelope.

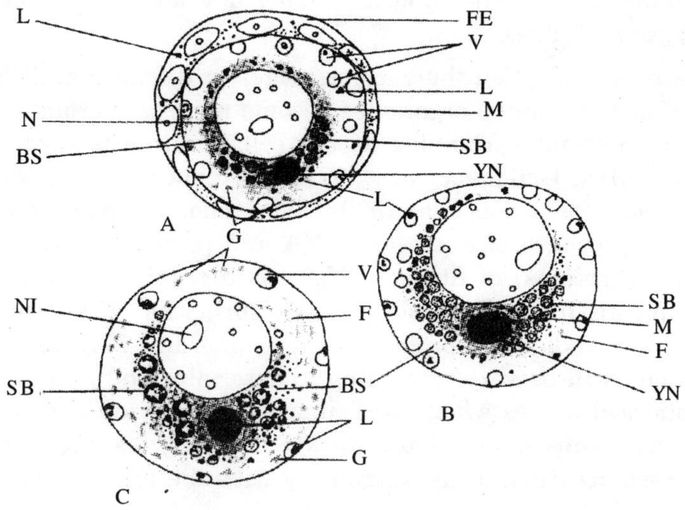

Fig. 4

The round or irregular mitochondria having the phospholipid-protein composition lie in association with the granular basophilic substance (Figs. 5A, 5B and 6A, 6B) (Hertig and Adams, 1967; Guraya, 1967a, 1970, 1974). These authors have reported a close morphological association of mitochondria with ER or granular basophilic substance. The mitochondria show the usual ultrastructure by having double membrane with arched cristae (Hertig and Adams, 1967; Hertig, 1968; Guraya, 1974). They are arranged in layers around the nucleus, in a preferential eccentrical distribution, particularly in oocytes at leptotene, zygotene and pachytene stages (Stegner and Wartenberg, 1963; Stegner, 1967; Baker and Franchi, 1967a). The mitochondria may also lie in association with electron-dense intermitochondrial substance or nuage, which is believed to consist of mRNA (Dvorak and Tesarik, 1980). The inner mitochondrial profiles show variations in their development and localization in different mitochondria of the same oocyte suggesting their different metabolic activity in relation to production of energy. The causative factors of these variations are required to be determined at the molecular level.

The spherical bodies stain homogeneously for RNA, proteins and lipoproteins in the primordial oocytes of young women in their early years (e.g., in 20s) of reproductive life (Fig. 4A and 4B) (Guraya, 1967a, 1970, 1974, 1999a, 2000a). Ultrastructurally the individual spherical body of such primordial oocytes shows a composite structure, consisting of small spherical granules or vesicles embedded in matrix of granular material (Figs. 5A, 5B and 6A, 6B). Such heterogeneous spherical bodies have not been observed in the oogonia and oocytes of foetal premature born, neonatal and early postnatal specimens (Fig. 5D) (Stegner and Wartenberg, 1963; Lanzavecchia and Mangioni, 1964; Baker and Franchi, 1967a, Makabe *et al.*, 1989, 1991; Wartenberg, 1989; Motta *et al.*, 1995, 1997a, b). Actually spherical bodies form the characteristic feature of the ooplasm in the primordial follicles of adult women (Fig. 4 (A-C)) (Guraya, 1970, 1974, 1999a, 2000a). They lie among the other ooplasmic structures of Balbiani's vitelline body and are smaller in size than the yolk nucleus or cytocentrum proper but larger than the other cell organelles. Spherical bodies show much variation in their number, size, morphology and histochemistry in women of varying ages

(Fig. 4 (A-C)). This indicates that the spherical bodies develop in preadolescent years of girls as suggested by (Guraya, 1999a).

Electron microscopic studies on the various stages in the normal differentiation of germ cells have revealed that the various organelles become more numerous and meanwhile their internal structure becomes more complex as the cells enlarge toward the diplotene stage (Baker and Franchi, 1967a; Gondos *et al.*, 1971; Wartenberg, 1989; Makabe *et al.*, 1989; 1991; Motta *et al.*, 1995, 1997a; Guraya, 1999a). The mode of formation of spherical bodies is not known. But the presence of abundant RNA and proteins in them suggests that the substances (RNAs and proteins) of nuclear origin as a result of transcription of maternal genes may be playing an important role in their formation. The application of modern immunocytochemical and molecular probes to the primordial follicles in various postnatal and preadolescent years will be very rewarding to determine the precise mode of origin and differentiation of spherical bodies which are found to be very sensitive to the age of women for their formation and subsequent deterioration or decay as they undergo conspicuous changes in their ultrastructure and histochemistry with the age of women (Guraya, 1999a).

In the ovaries of women in late 20s, the spherical bodies begin to develop sudanophilic lipid granules consisting of phospholipids (Fig. 4B) (Guraya, 1970, 1974, 1999a). This shows that with aging, the granules or vesicles of spherical bodies demonstrated with electron microscopy (Hertig and Adams, 1967; Hertig, 1968) start to undergo fatty change through the development of some more free lipids in them. From 30 years onward, more lipid granules of variable size begin to appear in the spherical bodies, and simultaneously they also show an increase in their size as well as in lipid content (Fig. 4C). Such changes become more pronounced in the primordial oocytes of women ranging in age from early 30s to 40 years. With aging of women, the granules or vesicles of spherical bodies originally filled with electron-opaque ribonucleoproteins develop triglycerides and some phospholipids; simultaneously they increase in size (Guraya, 1970, 1974, 1999a, 2000a). Spherical bodies containing lipid granules of variable size correspond to the large, compact, compound aggregates reported by electron microscopy (Figs. 5A, 5B, 5C and 6A, B) (Hertig and Adams, 1967; Hertig, 1968). These authors could demonstrate compound aggregates corresponding to spherical bodies of histochemical preparations (Guraya, 1970, 1974, 1999a) because they also used ovaries for the ultrastructural studies of primordial oocytes from women 28-37 years of age. In their electron microscopic studies, the other workers did not report comparable compound aggregates as they mostly used oocytes of foetal prematurely born, neonatal and early postnatal specimens (Stegner and Wartenberg, 1963; Lanzavecchia and Mangioni, 1964; Stegner, 1967; Makabe *et al.*, 1989; 1991; Wartenberg, 1989; Motta *et al.*, 1995, 1997a); in their morphometric comparative study of healthy and atretic human primordial and primary follicles, Stankova *et al.* (1985) have also not reported compound aggregates which form the characteristic feature of human oocytes.

The lipid granules of spherical bodies of histochemical studies (Guraya, 1970) are clearly visible as electron-transparent, spherical structures of variable size embedded in a matrix of compound aggregate enclosed by a membrane (Figs. 5C and 6A, 6B) (Hertig and Adams, 1967; Hertig, 1968). The matrix continues to stain mainly for RNA, protein and some lipoproteins. A close examination of large compound aggregates in electron micrographs published by Hertig and Adams (1867) shows the fusion of some small vesicles containing finely divided electron-opaque material.

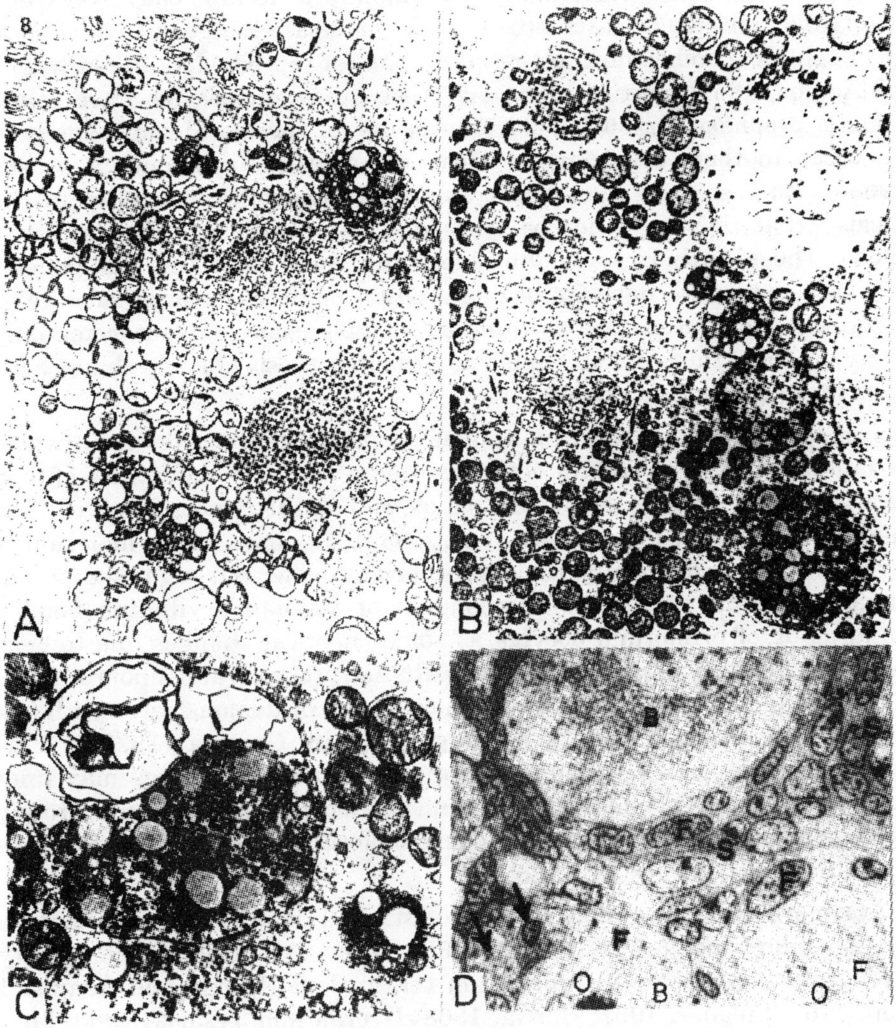

Fig. 5

With increase in lipid contents in women ranging in age from 30 to 40 years or more, the spherical bodies in most primordial oocytes show a progressive increase in disorganization or deterioration in their basic structure as they develop clear vacuoles of variable size (Fig. 4C) (Guraya, 1970, 1999a, 2000a) making the paranuclear cytoplasm of primordial oocytes frothy in appearance (Fig. 4C). These vacuolated spherical bodies correspond to the "ballooned compound aggregates" (Fig. 5C) of Hertig and Adams (1967) and Hertig (1968). During this vacuolization of spherical bodies, most of the basic matrix consisting of RNA, proteins and some lipoproteins disappears from view, their lipid granules composed mainly of triglycerides and some phospholipids form masses of variable size, which lie in association with the vacuoles of spherical bodies (Fig. 4C) (Guraya, 1970, 1999a) or "ballooned compound aggregates" (Fig. 5C) (Hertig and Adams, 1967; Hertig, 1968). The increase in the volume fraction of smooth ER and vacuoles as claimed by Stankova *et al.* (1985) for the so-called atretic primordial follicles of human ovary can be attributed to the vacuolization of compound aggregates or spherical bodies, which makes the Balbiani's vitelline body frothy in appearance (Fig. 4C). Actually various ultrastructural and chemical changes result in the distorted appearance of spherical bodies during the later years of reproductive life of women (Fig. 5C) (Hertig and Adams, 1967; Hertig, 1968). As these quiescent oocytes have remained in an arrested meiotic prophase for 30 to 40 years or more, therefore, some aging changes are expected to occur in them resulting in decay of ooplasmic components as suggested by Guraya (1999a). But the lack of demonstrable acid phosphatase activity in the spherical bodies indicates that they may not be lysosomal in nature as suggested by Hertig and Adams (1967). These authors have, however, suggested that the "ballooned compound aggregates" represent waste deposits of degradation or decay of organelles (e.g., spherical bodies, of histochemical preparations) similar to lipofuscin pigment commonly found in aging cells (Saver, 1984) thus further supporting the hypothesis of aging changes in primordial oocyte as put forth by Guraya (1999a).

Modern molecular probes must be applied to primordial oocytes from women in their middle and late reproductive years to determine more precisely the nature of aging changes in the spherical bodies at the molecular levels as well as the factors involved in causing the aging changes in their molecules (Fauser *et al.,* 1999). The use of such molecular probes will also be very rewarding to determine the precise origin, molecular organization and chemical changes of various other ooplasmic components of Balbiani's vitelline body during the formation and aging of primordial oocytes as some chemical changes due to decay are also expected to occur in them especially in their RNAs, proteins and lipoproteins during the later years of reproductive life of women, which could not be demonstrated with histochemical techniques and electron microscopy as suggested by Guraya (1999a). For example, mitochondria forming the source of energy for the primordial oocytes may undergo aging changes in their lipoproteins, DNA, RNA, and structural and enzymic proteins, for the demonstration of which the application of modern molecular probes will be rewarding (Fauser *et al.,* 1999; Satav and Nair, 1999).

Human primordial oocytes show masses of stacked or concentric annulate lamellae in the Balbiani's vitelline body (Figs. 5A and 6B). They are either in continuity with the nuclear envelope or occur some distance from the nucleus (Hertig and Adams, 1967; Baca and Zamboni, 1967; Stegner, 1967; Hertig, 1968; Dvorak and Tesarik, 1980; Makabe *et al.*, 1991; Motta *et al.*, 1995, 1997a, Guraya, 1974, 1999a). Other workers have not paid much attention to them in the human oocytes. Under the light microscope annulate lamellae appear as elongated and curved bodies consisting of RNA, proteins and some lipoproteins (Guraya, 1970, 1974, 1999a). Annulate lamellae representing specialized form of ER are known to be formed by fusion of vesicles originating from the blebbing activity of the two leaflets of the nuclear envelope (see Guraya, 1974, Kessel, 1985). Subsequently they increase in extension by fusion with the linearly arranged vesicles usually present at either end of each lamella. The pores in annulate lamellae are believed to be important in the release of stored informational macromolecules consisting of RNAs and proteins into the cytoplasm (Kessel, 1985) which are produced as a result of gene expression during oogenesis (Bachvarova, 1985; Fauser *et al* 1999). The involvement of annulate lamellae in the mobilization of stored gene products or with the processing of long-lived developmental information is required to be extended and confirmed with modern molecular probes (Fauser *et al.,* 1999). Baca and Zamboni (1967) have reported numerous flattened cisternae, which are lined by paired membranes seen at their ends. These cisternae have not been described by other workers (Hertig and Adams, 1967; Stegner, 1967; Hertig, 1968) who have, however, reported the presence of packed spiral fibrils, microtubules and vesicular aggregates.

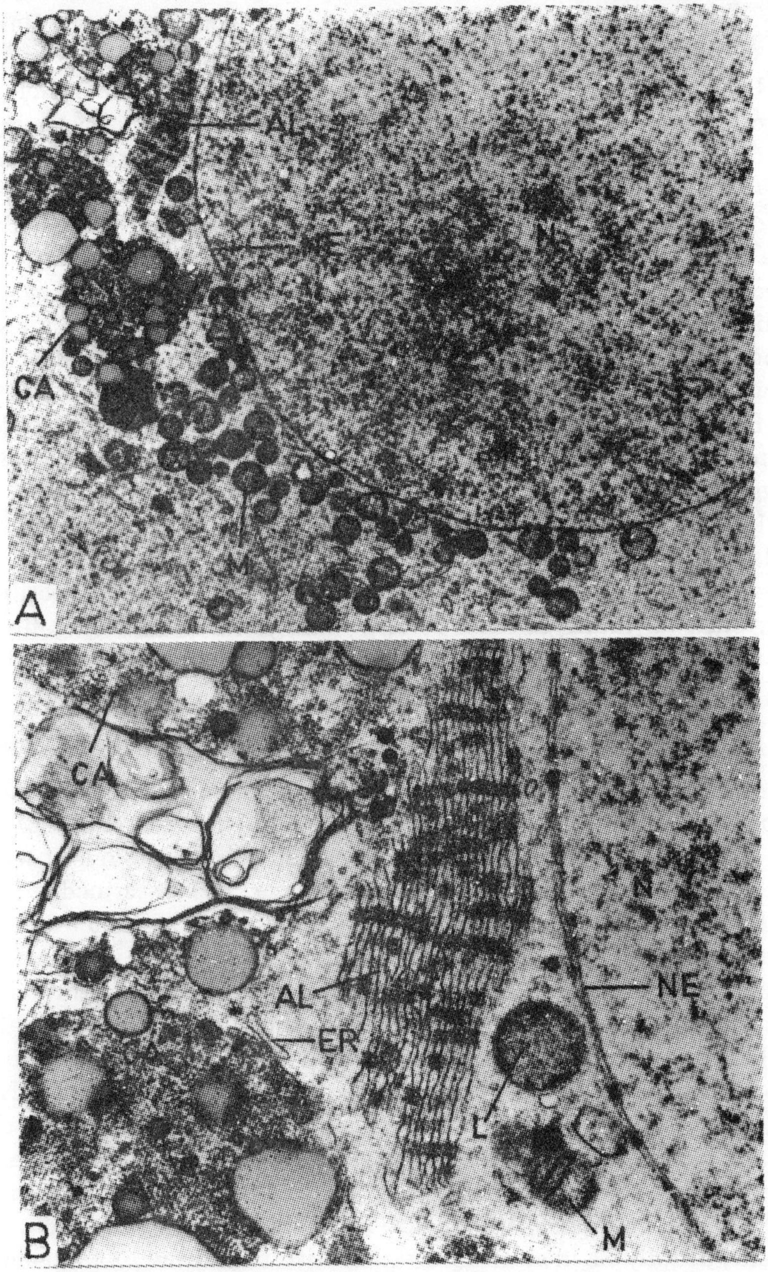

Fig. 6

Deeply sudanophilic lipid bodies of variable size, which consist of phospholipids of specific nature, occur in the ooplasm of primordial oocytes (Figs. 3A and B, 4 A-C) (Guraya, 1967a, 1970, 1974, 1985, 1999a, 2000a). A careful examination of lipid bodies under higher magnification of light microscope reveals that the heterogeneous structure of some of them is due to their sudanophilic and sudanophobic parts, some of them appear to be in the form of an aggregation of small lipid granules (Guraya, 1970, 1985). The lipid bodies lie in association with the cytocentrum and other cytoplasmic components of Balbiani's vitelline body (Figs. 3A, 3B and 4A-C). Some of them are also present in the peripheral ooplasm adjacent to the plasma membrane, where they generally form a close morphological association with the pinocytotic vacuoles (Figs. 3A, 3B and 4A-C). The number of both vacuoles and lipid bodies varies in the peripheral ooplasm of different primordial oocytes (Guraya, 1967a, 1970, 1974, 1985, 1999a, 2000a). Ultimately, both the components are dissociated from the plasma membrane of oocyte and are apparently shifted to the paranuclear ooplasm; large vacuoles generally merge with the ground cytoplasm. The sites of the distribution of lipid bodies clearly show that they are transported from the follicular or granulosa cells containing similar lipid bodies as suggested by their association with endocytotic vacuoles lying adjacent to the plasma membrane of primordial oocytes. Lipid bodies have been described as smaller, compact aggregates, usually without an enclosing membrane in ultrastructural studies of human oocytes (Hertig and Adams, 1967; Hertig, 1968). This is further supported by the fact that according to Hertig and Adams (1967), the vacuoles containing elements similar to those in the aggregate are present in the peripheral cytoplasm of oocyte. According to Guraya (1970, 1999a), they are clearly identical to the lipid bodies and vacuoles of the peripheral ooplasm observed in histochemical studies. Other workers using electron microscopy appear to have either overlooked the lipid bodies, or described them under different names, such as multivesicular bodies or organelles enclosing highly osmiophilic granular material (Stegner and Wartenberg, 1963; Lanzavecchia and Mangioni, 1964; Stegner, 1967; Baca and Zamboni, 1967; Baker and Franchi, 1967a).

1.3.2 Nucleus and Formation of Ribonucleoproteins and Other Components

The nucleus of the primordial oocyte is large and usually spherical and shows a large irregularly shaped nucleolus which consists of RNA and protein (Figs. 3A, 3B and 4A-C) (Guraya, 1967a, 1970, 1999a, 2000a); an incomplete shell of lipoprotein is observed to surround this large nucleolus. Electron microscopic studies have also described a large irregularly shaped nucleolus with a nucleolonema (Hertig and Adams, 1967). Baker and Franchi (1967a) have reported large nucleoli, which are uniformly granular but may contain clear, vacuole-like areas. According to Baca and Zamboni (1967) one or more dense, threadlike nucleoli are usually present. All these workers, using electron microscopy have not differentiated the lipoprotein shell around the large nucleolus that was observed by Guraya (1970) in histochemical preparations. Small nucleolus-like bodies consisting of RNA and protein (Guraya, 1967a, 1970, 1974, 1999a, 2000a), which are distributed around the periphery of the nucleus or are placed near the centre (Figs. 4A-C), correspond to the multiple, large clumps of fine granular material interpreted as heterochromatin by Hertig and Adams (1967). But Baker and Franchi (1967a) observed that smaller nucleoli, which are also uniformly granular, usually lie in close association with the chromosomal sheaths, suggesting their origin from the chromosomes possibly as a result of expression of ribosomal genes (Guraya, 2000a), which is required to be investigated with modern molecular probes for human (Fauser *et al.*, 1999; Yen *et al.*, 1999; Tevelde *et al.*, 2000; Bulletti *et al.*, 2001).

The nucleus of primordial oocytes shows filamentous chromatin which is uniformly distributed throughout the nucleus and stains for DNA and protein (Figs. 6A and 6B)

(Guraya, 1967a, 1970, 1974, 1999a, 2000a). DNA is now known to be synthesized or replicated just before the onset of meiotic prophase and continues to persist as no DNA is synthesized after last premeiotic interphase (Guraya, 1974, 1998a, 1999a, 2000a; Fauser *et al.*, 1999). According to Hertig and Adams (1967), aggregates of filamentous and granular chromatin are distributed throughout the nucleus in flocculent strands and apparently represent sections through the diplotene configuration of chromosomes (Figs. 6A and 6B). But Baca and Zamboni (1967) have stated that the fine granular chromatin is dispersed uniformly throughout the nucleus. But these workers appear to have made no distinction between the perichromatin or interchromatin granules and aggregates of filamentous chromatin reported by Hertig and Adams (1967).

Various ultrastructural studies have demonstrated that primordial oocytes in human remain at the diplotene stage. The chromosomes consist of DNA and protein whether at the diplotene or dictyate stage, they are organized in a fundamentally similar way and develop decondensed configuration (Baker and Franchi, 1967b; Guraya, 1974, 1999a). Their decondensed organization is supported by their structural components (1) an axis or core containing at least two longitudinal strands about 200A thick, (2) a surrounding sheath composed of coiled fibrils which form symmetrically arranged columns and loops and (3) clusters of large granules forming an association with the outer parts of the sheath. The apparent structural differences are believed to reflect variations in the compactness of axes and lateral loops of a decondensed type of organization. Fine granulation consisting of RNA and protein (Guraya, 1970, 1974, 1999a) apparently corresponds to the clusters of granules described by Baker and Franchi (1967a, b), and to the small perichromatin or interchromatin granules described by Hertig and Adams (1967). Autoradiographic studies have shown that the decondensed chromosomes of primordial oocytes form sites of increased synthetic activity as they rapidly incorporate radioactive precursors of RNA and protein (see Guraya, 1974, 1999a, 2000a; Fauser *et al.*, 1999; Yen *et al.*, 1999; Tevelde *et al.*, 2000; Bulletti *et al.*, 2001).

The nuclear envelope, which separates the nucleoplasm from the ooplasm usually consists of two membranes (Figs. 6A, 6B). It shows annules or pores (Hertig and Adams, 1967: Stegner, 1967: Baker and Franchi, 1967a; Hertig, 1968; Guraya, 1974, 1999a, 2000a). Baker and Franchi (1969) have reported the presence of dilations or "pockets" of variable size, which form between the two nuclear membranes in cells from the preleptotene stage of meiosis, thus forming a specific feature of oocytes at particular stages of differentiation. From the formation and morphological variations of dilations at different stages of meiosis, Baker and Franchi (1969) have implicated the nuclear envelope in the formation of cytoplasmic inclusions (or organelles) during the early growth and differentiation of cytoplasm of the oocyte (preleptotene to early diplotene), when there occurs a two-fold increase in the cytoplasmic/nuclear ratio of the oocyte. The complex dilations are most frequently seen in this intermediate growth period, when the number of mitochondria and the complexity of other cytoplasmic organelles show a marked increase (Hertig and Adams, 1967; Guraya, 1970, 1974, 1999a; Fauser *et al.*, 1999; Yen *et al.*, 1999; Tevelde *et al.*, 2000; Bulletti *et al.*, 2001). According to Baker and Franchi (1969) inclusions seen within the dilations may be a previous stage in the production of the "mitochondrial precursor bodies" described by Wartenberg and Stegner (1960) in the cytoplasm of oocytes in postnatal human ovaries. The dilations of the nuclear envelope are not observed after the diplotene stage is established and primordial follicle is completed (Baker and Franchi, 1969). This may explain why various workers (e.g., Wartenberg and Stegner, 1960; Hertig and Adams, 1967; Baca and Zamboni, 1967) have not observed these structures in the oocytes of prepubertal and mature girls.

Various RNAs synthesized by the decondensed chromosomes and nucleolar rRNA (Fauser *et al.*, 1999) are transported into the ooplasm through the pores or annuli or nuclear envelope. Some of the RNA must be utilized as messenger for protein synthesis by the ribosomes in the primordial oocytes, but major portion of messenger RNA and rRNA must be stored in the ooplasm for some use during oocyte growth (Guraya, 1974, 1999a, 2000a; Fauser *et al.*, 1999, Chapter 2). The signals, mechanisms and regulation involved in the differential transcription of genes of various RNA species (i.e., rRNA, mRNA, and tRNA) and their subsequent transport into the ooplasm through the pore complexes of nuclear envelope remain to be investigated at the molecular level during the formation and aging of human primordial oocytes (Fauser *et al.*, 1999; Bulletti *et al.*, 2001). Even the chemical changes of various RNAs and proteins in the quiescent primordial oocytes form promising area for future investigations with modern molecular probes (Fauser *et al.*, 1999). Verrotti and Strickland (1997) have reported oocyte selection of mutations affecting cytoplasmic polyadenylation of maternal mRNAs. The integration of DNA replication and RNA biogenesis in the cell nucleus, and protein synthesis in the cytoplasm of eukaryotic organisms is now known to depend on selective transport of proteins and ribonucleoproteins between these two compartments (Nigg, 1997; Fauser *et al.*, 1999). The study of effects of aging on the nucleocytoplasmic transport mechanisms at the molecular level will be very rewarding to determine more precisely the nature of aging changes in various informational macromolecules, such as mRNA, rRNA, tRNA, DNA, proteins etc. of human primordial oocytes and their subsequent effects on oocyte growth and maturation and, embryogenesis in relation to various human infant disorders.

1.3.3 Follicle Wall

Primordial oocytes get invested by a layer of flattened and/or polyhedral cells which derive from the undifferentiated stromal cells and correspond to the future follicular or granulosa cells (Makabe *et al.*, 1989, 1991; Wartenberg, 1989; Motta *et al.*, 1995a, b, 1997a, b). The primordial follicle wall forms a crescent-shaped cup of relatively large cells over parts of the oocyte (Guraya, 1970; Motta *et al.*, 1994); the rest of the follicular or granulosa cells form a very thin layer around the rest of the oocyte (Figs. 3A, 3B and 4A). A basal lamina lies between the follicular cells and surrounding stromal cells as already stated. The elongated and highly irregular nucleus of undifferentiated follicular cells due to the presence of numerous folds and profound indentations shows one or two reticular nucleoli consisting of RNA and proteins, heterochromatin masses composed of DNA and proteins, and a few pores in the nuclear envelope. The rod-shaped, elongated and tortuous mitochondria having usual lipoprotein composition, numerous transverse cristae and an electron-dense matrix, lie especially in the widened parts of follicular cells. The phospholipid bodies, a very small Golgi complex, a few randomly distributed elements of granular ER and free ribosomes are present in the cytoplasm of follicular cells, polysomes are numerous (Guraya, 1970, 1971, 1999a, 2000a; Makabe *et al.*, 1991; Motta *et al.*, 1995, 1997a). All these cytological and histochemical features are indicative of synthesis of some proteins in the follicular cells, which remain to be localized immunocytochemically, and isolated and chemically characterized for determining their autocrine and paracrine functions in the quiescent primordial follicles. According to Bennett *et al.* (1996) immunohistochemical staining of transforming growth factor a, epidermal growth factor (EGF) and EGF receptor was significantly more intense in the oocytes than in the stroma of the human foetal ovary. These results combined with determination of mRNA for these growth factors suggest a role for them in the human foetal ovarian development. The strong staining in primordial oocytes suggests a possible autocrine or paracrine role of these growth factors in the human oocyte growth and differentiation in utero. Danforth (1995) has also suggested the paracrine control of oocyte development in the human.

Frequently lipid bodies consisting of phospholipids are observed lying across the plasma membrane of the primordial oocytes where they are associated with vacuoles possibly formed by pinocytosis (Figs. 3A, 3B and 4A) as already stated. The lipid bodies consisting of phospholipids correspond to the small compound aggregates observed in the follicle cells in electron microscopic studies, which have also been observed to traverse the oocyte membrane and are often associated with vacuoles in the oocyte cytoplasm (Hertig and Adams, 1967: Hertig, 1968). Secondary lysosomes, are observed in the follicle cells (Dvorik and Tesarik, 1980). The spherical bodies forming the characteristic feature of primordial oocytes and showing aging changes are not developed in their follicular cells suggesting that they play some important role in the developmental process of oocyte, which is required to be determined at the molecular level.

1.3.4 Structural Relationships Between the Follicular Cells and the Oocyte

The final structural relationship between the follicular cells and the oocytes in human primordial follicles can be characterized by variable patterns of association (Stegner and Wartenberg, 1963: Stegner, 1967: Lanzavecchia and Mangioni, 1964: Makabe *et al.*, 1989, 1991; Wartenberg, 1989; Motta *et al.*, 1994, 1995, 1997a). For most of the extension, the plasma membrane of the primordial oocyte is smooth and directly apposed to the surrounding flattened and/or polyhedral follicular cells which may project microvillous processes in some areas (Motta *et al.*, 1994, 1995, 1997a; Guraya, 1999a, 2000a). These are in apposition with the oolemma and form bulbous terminals presenting attachment zones or zonulae adherentes, desmosomes and communicating (gap) junctions, which are characterized essentially by increased opacity of the plasma membranes. Similar cell attachments are also frequently observed between adjacent follicular cells and their microvillous extensions. Occasionally, the oocyte and follicle cells are separated by narrow extracellular space in the form of irregular slits. In these zones the plasma membrane of the oocyte is extended into a few projections or microvilli which may indent the cytoplasm of follicle cells. The microvilli are generally most prominent between adjacent follicle cells in the follicle oocyte junction.

The membranes of follicle cells and oocyte may also be locally joined by desmosomes characterized by increased thickness and electron opacity of the plasma membranes and convergence of cytoplasmic filaments onto the attachment areas or they may be separated by irregular lacunae; in this case, the plasma membrane of either cell forms short interdigitating microvilli and the surface of the follicular cells. Numerous gap junctions develop between the oocyte and the follicular cells in response to the increasing metabolic activity by the oocyte, which gets excluded from the vascular compartment during the formation of primordial follicle (Motta *et al.*, 1995, 1997a). These junctions play an important role in the maintenance of oocyte by facilitating the incorporation of nutrients and favoring the synchronous differentiation of the oocyte-follicular cell complex.

The associative patterns between adjacently placed follicular cells also vary as in some places their membranes are closely apposed against one another, and are frequently associated through interposition of tight intercellular junctions (or desmosomes). Substances of metabolic significance possibly reach the oocyte after passing through the cytoplasm of follicular cells by common methods of transcellular transport. The movement of molecules across follicular cell body is potentially a more selective process involving receptors, pumps exchange mechanisms, enzyme systems, transport and internalization systems as suggested for Sertoli cells (Guraya, 1999b). Whatever route is followed, it appears that the substances enter the primordial oocyte through processes of membrane diffusion, pinocytosis and active transport (Guraya, 1974, 1985, 1998a, 1999a, 2000a). The study of effects of aging on various transport devices/mechanisms involved in the transfer of nutrients, metabolic and regulatory substances in human primordial follicles form a promising area for research with modern immunocytochemical and molecular techniques. The results of such studies will also help us to know whether the changes in transport mechanisms at the molecular level within the primordial follicles lead to aging of their ooplasmic components.

1.3.5 Stromal Cells

Surrounding stromal tissue lying outer to the basal lamina consists of spindle-shaped fibroblast-like cells (Figs. 3A, 3B and 5D). Histochemical studies have demonstrated the presence of RNA, proteins and phospholipid bodies similar to those of follicle cells (Guraya, 1967a, 1970, 1998, 1999a, 2000a). The electron microscopic studies have shown the presence of elements of rough ER, small Golgi complex, free ribosomes and mitochondria with simple cristae and small compound aggregates (Hertig and Adams, 1967; Hertig, 1968). The phospholipid bodies of histochemical studies correspond to the ultrastructural compound aggregates. There is seen a continuity between the phospholipid bodies of oocyte, follicle cells and surrounding stromal tissue (Figs. 3A, 3B) suggesting their some specific metabolic functions in primordial follicles which are required to be determined at the molecular level. The precise roles of stromal cells in the regulation of initiation of growth, differentiation and aging processes of human primordial oocytes remain to be determined at the molecular levels. In this regard some growth promoting and inhibiting factors of polypeptide nature may be playing significant roles in the processes of intercellular communication between various components of the primordial follicles and surrounding stroma as suggested for the testis (Guraya, 1998a, 1999b) which remain to be isolated, identified and chemically characterized in both *in vitro* and *in vivo* systems (Adashi, 1990; Gougeon, 1996a, b).

1.4 ALTERATIONS OF OVARIAN PRIMORDIAL FOLLICLES, STROMA, BLOOD VESSELS ETC. WITH AGING OF WOMEN

The human ovary is well known to have abundant stromal tissue and a rich vascularity to facilitate follicular growth, the production of steroid hormones and polypeptide growth factors and its sensitivity to hormonal agents (Guraya, 1985, 1997a, 2000a; Shimada *et al.*, 1993; Yen *et al.*, 1999; Fauser *et al.*, 1999; Tevelde *et al.*, 2000; Bulletti *et al.*, 2001). Both the ovarian components revealed conspicuous morphological and histochemical changes with the aging of women (Figs. 7A, 7B; 8A, 8B). In ovarian tissues from to 30-year-old woman, a few spiral arteries in the hilus and medulla showed a normal appearance, but most exhibited a thickned tunica intima and splitting of the internal elastic lamina (Shimada *et al.*, 1993). But in the 38-43 year old group arteries in both the hilus and medulla showed a considerable thickening of the tunica intima. Guraya (1976b) made a study of correlative histochemical alterations of ovarian stroma and blood vessels with the advancing age of women, 18 to 52 years of age. The fatty metamorphosis of stromal tissue in localized areas of the ovarian cortex (Figs. 7A, 7B), which are identical to stromal lipid bands of Fienberg

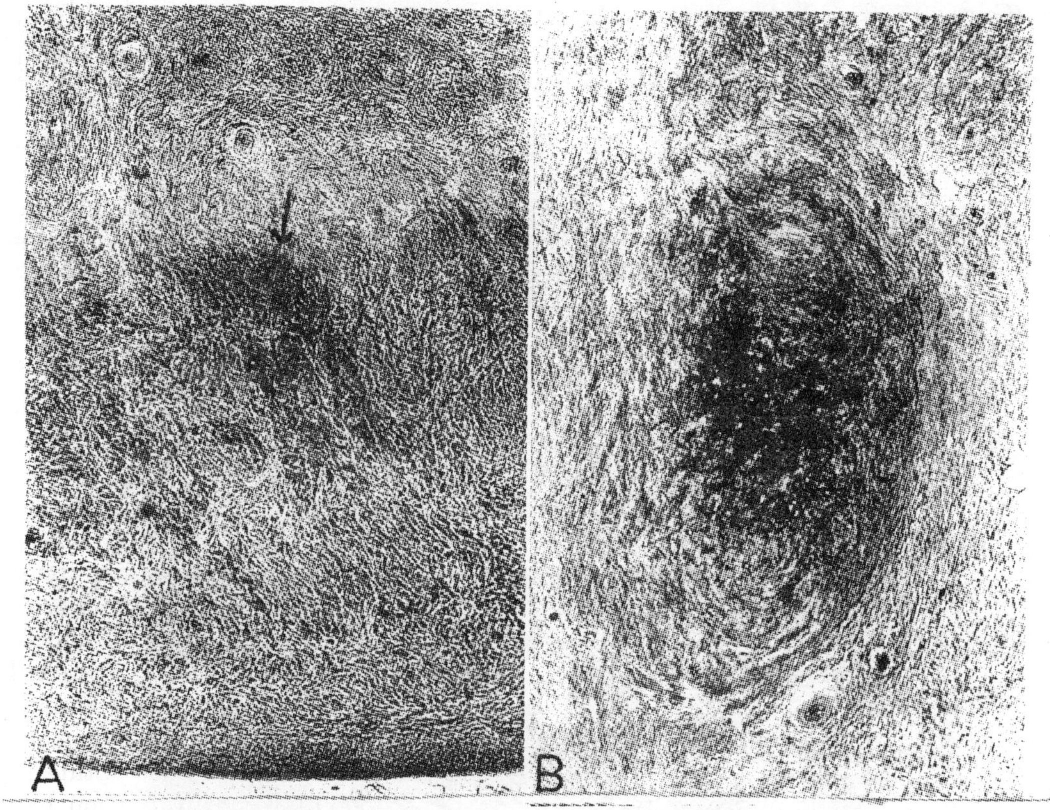

Fig. 7 Portion of ovarian medull from the postmenopausal of 45 years, showing the appearance of some sundanophilic lipids in the endothelial lining of blood vessels. × 96.

and Cohen (1965), begins to appear in the third decade and in subsequent decades it progressively increases to form a conspicuous feature of human ovary (Fig. 8A). It is closely accompanied by the development of lipids in patches, which consist of first phospholipids (in the third decade) and then triglycerides, phospholipids and some cholesterol (in the fourth and fifth decades). Correspondingly there also occur similar lipid changes in the wall of blood vessels of ovarian medulla (Fig. 8B). Guraya (1976b) suggested that there exists a close association or correspondence in the aging changes of primordial oocytes, cortical stroma and blood vessels of human ovary with advancing age. Further studies using a variety of modern molecular and physiological techniques should be carried out to understand the molecular, sub-cellular and physiological aspects of alterations in the cortical ovarian stroma and medullary blood vessels and their subsequent implications in the aging process of primordial oocytes (Fauser *et al.*, 1999; Bulletti *et al.*, 2001). The gaps in our knowledge of these relationships form very challenging problems to be filled in future investigations with multidisciplinary approach for ensuring the development, maturation and ovulation of quality oocytes for normal embryogenesis. The aging process is believed to be regulated by dual mechanisms. One is the mechanism which programs ultimate cell death; and therefore determines the life span at birth. The other is mechanism, which causes aging stochastically by damaging certain substances in tissues/cells with time in life as described for primordial oocytes (Guraya, 1999a; Section 3). Now these mechanisms are being investigated on genetical basis (Suzuka, 1993; Bulletti *et al.*, 2001). The important roles of oxygen free radicals in the process of aging is also suggested, which are required to the investigated for primordial oocytes.

A

B

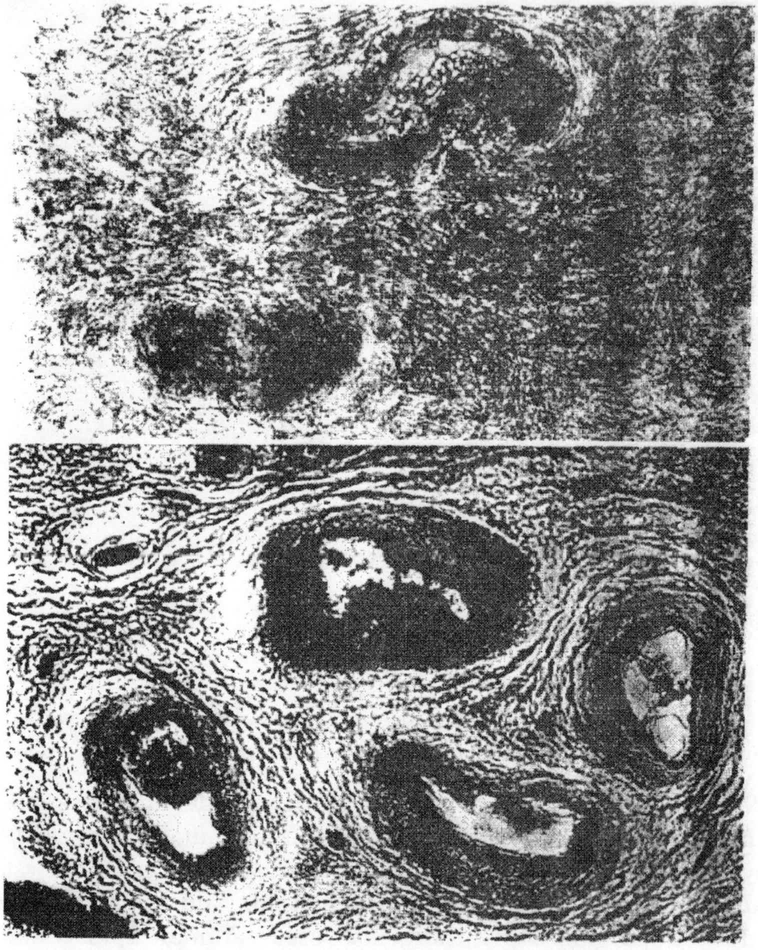

Fig. 8 (A) Portion of ovarian medulla from the post-menopausal woman of 45 years, showing the appearnace of some sub-danophilic lipids in the endothelial lining of blood vessels. × 96. (B) Portion of ovarian medulla from the post-menopausal woman of 52 years, showing heavy accumulation of deeply sub-danophilic lipids in the endothelial lining of blood vessels. × 96.

Some studies have been made of human ovarian aging and mitochondrial DNA (mDNA) deletion (Suganuma *et al.*, 1993; Bulletti *et al.*, 2001). The compact mDNA is believed to mutate because mDNA is very simple, not having histones or repairing systems and the exposition to oxygen radicals leaked from the electron transfer system. The mutations, such as point nucleotide mutation or deletion, will result in a defect in the respiratory complexes, leading to decreased energy production in the cell (see Suganuma *et al.*, 1993). Moreover, not only this process, mDNA deletion may cause oxidative stress with consequent peroxidation of membrane lipids. The impaired molecular assembly of the respiratory complexes due to mDNA mutations will increase the generation of oxygen free radicals in mitochondria, leading to further degeneration of cells. Such molecular changes in mDNA and their physiological/developmental effects are expected to occur in the primordial oocytes, which have been shown to undergo aging changes as already discussed in section 3.1. The effect of aging on energy production has attracted a lot of attention; and cumulating evidence shows a decrease in the activity of mitochondrial oxidative phosphorylation with increasing age (Suganuma *et al.*, 1993).As the primordial oocytes undergo aging changes, their function possibly also decreases, which is required to be investigated at the molecular level (Fauser *et al.*, 1999; Bulletti *et al.*, 2001). But there is a general relationship between the ovarian aging and the accumulation of mDNA mutation (Suganuma *et al.*, 1993). Since aging of primordial oocytes must be a very complex physiological process, various factors could be involved for the mechanism of their aging. The analysis of those factors will also be rewarding to resolve the clinical problems not only of menopause but also of developmental disorders of infants in aged women (Chapter 7).

2

Oocyte Growth

The growth and differentiation of oocyte during the follicular development are regulated by the activity of granulosa cells whose functions are regulated by various hormones and factors as discussed in detail by Guraya (2000a) in different species of mammals including primates and human (see also Osuga *et al.*, 1999; Yen *et al.*,1999; Fauser *et al.*, 1999;Tain *et al.*, 2000; Tevelde *et al.*, 2000; Sharma, 2000; Bulletti *et al.*, 2001). In other words, the oocyte growth is the result of interactions between the somatic cells and germ cells at the cellular and molecular level. Development of human preantral follicles has been studied *in vitro* (Abir *et al.*, 1999; Fauser *et al.*, 1999). Fluctuations in serum mullerian-inhibiting substance (MIS) levels during the menstrual cycle in human suggest that MIS may have a regulatory role in folliculogenesis (Cook *et al.*, 2000) which remains to be determined at the molecular level (Fauser *et al.*, 1999). Gentry *et al.* (2000) have suggested that the proteins involved in the thrombin-generating and chromatin modulatory pathways may be derived from ovarian cells, suggesting that thrombin may have a role in folliculogenesis in the human ovary, which is required to be determined at the molecular level (Fauser *et al.*, 1999). Follicular growth and maturation are mostly independent of genadotropins, from the stage of primordial follicles to autral follicles and a complete intraovarian paracrine system is implied in this gonodotropin independent follicular growth, and in the modulation of the actions of the gonadotropins. Various authors have discussed the endocrine, paracrine and autocrine regulation of the human menstrual cycle (see Guraya, 2000a). Zhao *et al.* (2000) have suggested that interfollicular factors are also involved in follicle development *in vitro*, which especially at the early folliculogenesis stage play a positive role in terms of follicular growth as well as survival. Hovatta *et al.* (1999) have studied the effect of partial isolation of human primordial, primary and secondary ovarian follicles in long term culture. The oocytes and follicles show considerable heterogeneity in structure, composition and organelle distributions over the course of oocyte growth and maturation (Guraya, 1974, 1985, 2000a; Gosden and Bownes, 1995; Fauser *et al.*, 1999). Such differences are of great physiological and developmental significance. The oocyte growth and maturation are the result of various cellular and molecular changes occurring during oogenesis as reviewed in detail for different species of mammals and human (Guraya, 1985, 2000a; Fauser *et al.*, 1999).

Actually the ability to mature, be fertilized and finally to develop into a viable embryo is acquired gradually by the oocyte during progressive differentiation throughout folliculogenesis

(Meremillod *et al.*, 1999; Fauser *et al.*, 1999; Wolf and Zelinski-Wooten, 2001). The effects of long term culture and partial isolation on human folliculogenesis are being studied (Hovatta *et al.*, 1999; Wolf and Zelinski-Wooten, 2001). A better understanding of the family of genes involved in oocyte and follicular atresia should allow a better understanding of premature ovarian failure (Fauser et al., 1999; Christin-Maitre and Beuchard 1999; Sharma, 2000). At present, hormonal replacement is given to these patients, suggesting oocyte donation.

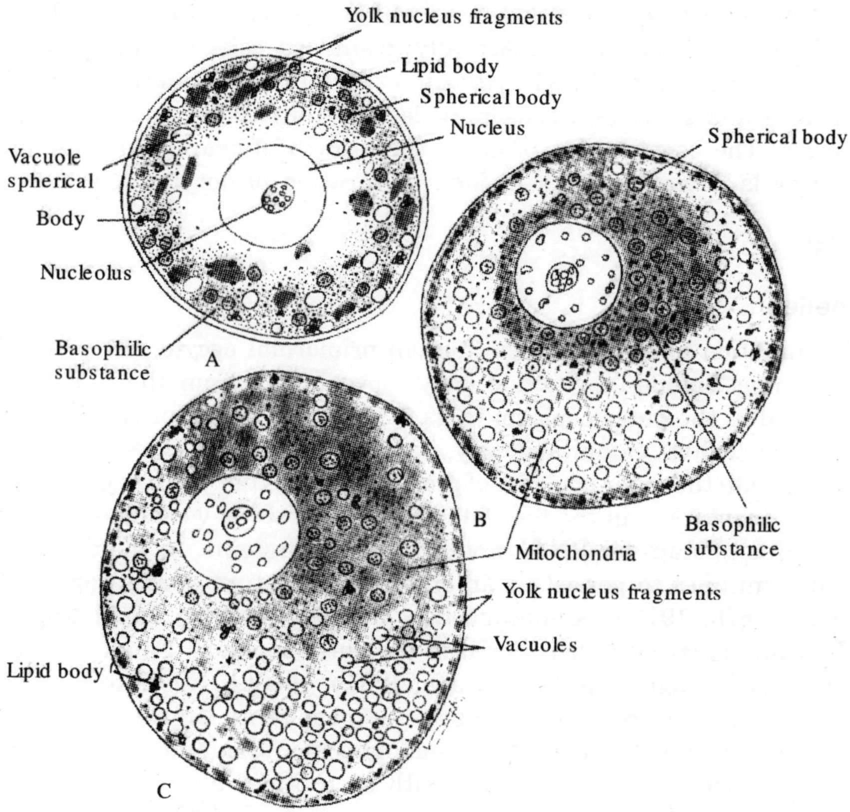

Fig. 9

Here emphasis will be laid on cellular and molecular aspects of oocyte growth, maturation and ovulation in human. A growth programme initiates and controls differentiation in the immature oocyte while a separate programme regulates reprogramming of the oocyte before ovulation (Fauser *et al.*, 1999; Guraya, 2000a) A better and through knowledge of human oogenesis is required to control the quality and fertility of oocytes for fertilization and early embryogenesis *in vivo* and *in vitro* (Wolf and Zelinski-Wooten, 2001). The oocyte at the start of its growth phase shows two features which include (1) its small size and (2) its total inability to progress from the G2 to the M phase of the meiotic cycle (Fauser *et al.*, 1999): Both these features are changed during the growth phase of oocyte, which after exit from the resting stage in the cortical stroma, begins to grow very fast and its volume is greatly increased by accumulating organelles, proteins, RNAs, lipids etc. (Fauser *et al.*, 1999) (figs. 9A, 9B) Approximately, half the amount of proteins required for this fast growth is produced by the oocyte itself and the remaining amount is definitely produced by the granulosa cells and then transported to the oocyte in a non-degraded form (Guraya, 1985, 2000a; Gosden and Bownes, 1995). The precise nature and amount of proteins contributed by the granulosa cells to the growing oocytes still remain to be determined. However, various nutrient substances required for the metabolic activities and various precursors needed for the synthesis and storage of informational molecules (RNAs and DNA) and proteins are transported from the granulosa cells by specific membrane specilizations of granulosa cells and oocyte as will be discussed (Section 2.3). The granulosa cells are also coupled to each other. Actually granulosa cells including cumulus and the oocyte form a gap junction-mediated syncytium. Actually the intercellular communication between the oocyte and granulosa cells provides the nutritional and metabolic support for the oocyte growth. The oocytes also promote the growth of granulosa cells and the growth promoting activity is decreased with meiotic maturation of oocytes (Guraya, 2000a).

2.1 OOPLASM

2. 1.1 Organelles

With the initiation of growth in the human primordial oocyte, the perinuclear complex of organelles forming Balbiani's vitelline body moves away from the nuclear envelope and becomes distributed in the outer ooplasm (Figs. 9A, 9B and 9C) (Guraya, 1961, 1967b, 1970b, 1972, 1985, 1998a,b, 2000a). Simultaneously, the yolk nucleus (or cytocentrum) fragments. As its fragments move into the outer ooplasm of growing oocyte, they further proliferate, apparently by growth and fragmentation, in the form of irregular masses of variable size and density (Fig. 9 A), which continue to stain for RNA, proteins and lipoproteins. In the oocytes of marmoset, the yolk nucleus continues to persist as an organized structure during some stages of oocyte growth (Guraya, 1967b, 1970b), simultaneously it seems to form small fragments from its periphery, which are apparently distributed in the outer ooplasm. Finally this persisting yolk nucleus also undergoes fragmentation (Guraya, 1967b, 1970b). The mitochondria staining for lipoproteins, and granular basophilic substance consisting of ribonucleoproteins (RNA, ribosomes, elements of granular and smooth endoplasmic reticulum) and phospholipid bodies, which lie in close morphological association with each other, also increase (Fig. 9A, 9B and 9C). The number of mitochondria increases, during oocyte growth as a result of blebbing of the nuclear envelope (Guraya, 1974, 1985, 2000a). They show great diversity in form and structure. Actually in the oocytes of primates, complex mitochondria are relatively common (Guraya, 1974). The physiological significance of these variations in the internal structure of mitochondria in the differentiating oocyte is still not known (Guraya, 1974, 2000a). But it will be rewarding to study if there occur any differences in the internal structure of mitochondria in the growing oocytes of young and aged women. In several places the mitochondria are closely associated with electron-dense material in the rhesus monkey oocyte, which resembles the mitochondrial rosettes or intermitochondrial substance described in the oocytes of other mammals (Guraya, 1970c, 1974, 1985, 1997a,b, 2000a). They are believed in some way associated with elaboration of the mitochondria.

The spherical bodies of human oocytes, described in Chapter 1 do not seem to increase in number during oocyte growth (Fig. 9A, 9B and 9C). These heterogeneous spherical bodies consisting of small granules or vesicles embedded in a matrix of fine granular material have also been reported in the developing human oocyte by other workers (Guraya, 1974), however,

Zamboni *et al.*, (1966), and Baca and Zamboni (1967) have not reported them in the growing and mature oocytes of humans.

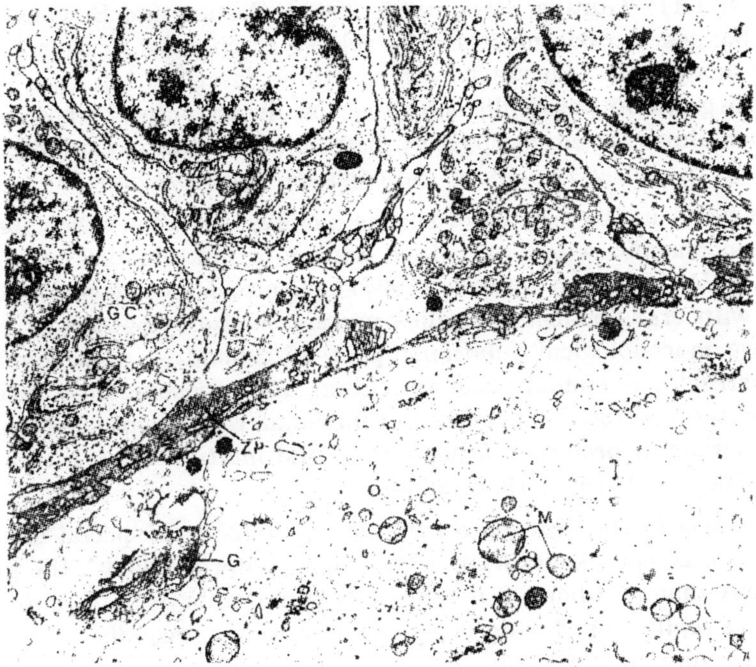

Fig. 10

In the growing oocytes of human, which are surrounded by two to three layers of granulosa cells. There is seen further accumulation of Golgi complexes, mitochondria, endoplasmic reticulum, ribosomes, phospholipid bodies etc., which are mainly distributed in the outer regions of the ooplasm leaving the perinuclear ooplasm entirely free of them (Figs. 10, 11) (Guraya, 1967b, 1970b, 1972, 1974, 1985, 2000a). The relatively large spherical masses derived from the yolk nucleus, which lie in the cortical ooplasm (Fig. 9A) seem to be the secondary sites for the growth and multiplication of yolk nucleus derivatives. When studied with the electron microscope, they are clearly identical to Golgi bodies, Golgi areas, Golgi complexes, or large aggregates of vesicles lying in the cortical ooplasm just beneath the plasma membrane of growing oocytes of primates and human (Baca and Zamboni, 1967; Guraya, 1974, 1985, 2000a; Gosden and Bownes, 1995). The peripheral Golgi bodies are concerned with the adjustment of fluid reserves in the oocyte and with the formation of zona pellucida as they lie just bellow the oocyte membrane, and in this regard it is of interest that in the oocytes of rhesus monkey they are frequently found adjacent to the newly formed spaces between the granulosa cell and the oocyte (Guraya, 1974). Baca and Zomboni (1967) found that the peripherally placed Golgi masses are involved in the formation of cortical granules as also reported by various other workers (Guraya, 1985, 1997, 2000a; Gosden and Bownes, 1995; Sharma and Chowdhury, 1998). The mitochondria and Golgi complexes are not only changed in number but also in structure during the oocyte growth, suggesting some functional and metabolic changes. The peripheral location of most organelles and phospholipid bodies suggests that the oocyte is equipped for the absorption, utilization, and intracellular transport of materials delivered to its surface by the surrounding granulosa cells constituting corona radiate cells as will be discussed in detail in Section 2.3.

The number of ribosomes present in polysomes is increased several fold during oocyte growth (Figs. 10 and 11) (Guraya, 1974, 1985, 2000a; Gosden and Bownes, 1995; Fauser *et al*., 1999). The changes in ribosomes may be related to changes in overall rates of protein synthesis during this period of oocyte growth as will be described later on. The mitochondria associated with the granular basophilic substance (or ultrastructural ribosomes and elements of endoplasmic reticulum) are either sparsely distributed or form irregular masses of variable size and density in the outer regious of the ooplasm (Figs. 9A and 11).

The mitochondria in the oocytes of rhesus monkey are often associated with fattened vesicles of the endoplasmic reticulum, which appear to link clusters of mitochondria into strands (Guraya, 1974). Baca and Zomboni (1967) have also drawn attention to the association of mitochondria with ergastoplasmic profiles present in the cytoplasm of human follicular oocytes. The close morphological relationship of mitochondria with the RNA-containing substance (ultrastructural endoplasmic reticulum and ribosomes) or intermitochondrial substance during oocyte growth suggests a functional relationship between them, which may be involved either in protein synthesis or in the multiplication of mitochondria or in both (Guraya, 1970c, 1972, 1974, 1985, 2000a).

The mitochondria in the growing oocyte show a number of special features of interest (Guraya, 1974, 1985, 2000a). Apart from a very small paternal "leak", the great majority of mitochondrial DNA that is inherited by the developing embryo is of maternal origin, and thus the fidelity of replication during oogenesis is critical, Secondly, the phenotype of the mitochondria is peculiar and possibly reflects attenuation of oxidative phosphorylation as source of energy (Satav and Nair, 1999). In the early stages of oocyte growth, mitochondria may be found in the vicinity of endoplasmic reticulum and have transformed to dumb-bell shapes (Guraya, 2000a). In the preovulatory oocytes, the number of small or oval or round mitochondria progressively increased although species differences are observed in regard to ooplasmic organelles (Guraya, 2000a). But it will be interesting and rewarding to determine the differences in the ooplasmic organelles of young and aged women.

The lipid bodies consisting of phospholipids also increase in number and size during oocyte growth (Figs. 9A, 9B and 9C) (Guraya, 1961, 1967b, 1970b. 1972, 1974, 1985, 1997a,b, 2000a). They show a heterogeneous structure because of their deeply sudanophilic and sudanophobic areas. They occur as fenestrated bodies of variable size. Most of the lipid bodies lie in the peripheral ooplasm adjacent to the plasma membrane of the growing oocyte, where some of them may be associated with pinocytotic vacuoles; the latter react negatively with the various histochemical tests used. The lipid bodies and vacuoles appear to dissociate from the plasma membrane and gradually move into the inner ooplasm, where they are distributed among the other ooplasmic components. The heterogeneous lipid bodies, which form the most prominent feature of growing human and primate oocytes, either have been overlooked in electron microscope studies or correspond to "vesicular conglomerates", multivesicular bodies, or electron-dense bodies (Guraya, 1974, 1985, 1997, 2000a; Gosden and Bownes, 1995).

During the growth phase, the diameters of oocytes increase while they are still arrested in meiotic prophase. During this period, not only the size of oocyte increases but also differentiations occur in regard to amount, structure, and distribution of various ooplasmic components (Figs. 9A, 9B and 9C) (Guraya, 1985, 2000a). It will be rewarding to make comparative studies of distributional, structural and chemical changes of various ooplasmic components in the oocytes of young and aged women (Bulletti *et al.*, 2001).

2.1.2 Cortical Granules

It is well established that the cortical granules and zona pellucida of mammalian and human eggs participate in a secondary block to polyspermy following their fertilization as will be discussed in Chapter 5 (Guraya, 1982, 2000b; Yanagimachi, 1994; Fauser *et al.*, 1999). They consist of carbohydrates and proteins. Corresponding to the accumulation of various ooplasmic components in the peripheral ooplasm of human oocyte is the development of PAS-positive granules (Figs. 12A, 12B) (Guraya, 1967b, 1969a, 1974, 1982, 1985, 2000a; Gosden and Bowness, 1995). They are formed by the activity of peripherally distributed Golgi masses of large size, multivesicular bodies and endoplasmic reticulum as demonstrated with correlative cytochemical, immunological and electron microscopic studies (Guraya, 1974, 1982, 1985, 2000a), which clearly correspond to the large yolk nucleus derivatives observed with histochemical studies (Guraya, 1967b, 1972, 1974, 1982, 1985, 2000a). The cortical granules lying in the peripheral ooplasm are membrane limited and contain relatively electron-dense, homogeneous material. The cortical granules of human oocytes are distributed as small granules adjacent to the plasma membrane (Fig. 11) and consist of carbohydrates and some proteins (Guraya, 1969a, 1974). At maturity, the cortical granules lie just beneath the oolemma (Fig. 11) to get ready for discharge in response to fertilization or other activating stimuli (Chapter 5). The cortical granules fuse with the oolemma at fertilization and by releasing their substances consisting of carbohydrates and proteins (glycoproteins) into the perivitelline space change the functional characteristics of the zona pellucida cause secondary block to polyspermy (Guraya, 2000b; Chapter 5).

2.1.3 Yolk

Developing oocytes of human and other primates, which greatly increase in volume at this stage, do not show the formation of yolk bodies. The large vacuoles (Figs. 9A, 9B and 9C), which do not develop sufficient material demonstrable with histochemical techniques (Guraya, 1967b, 1970b, 1972, 1974c), appear to correspond to the yolk vacuoles or lattices of other mammalian species (Guraya, 1985, 1987, 2000a). They appear to be filled with fluid. The exact source of their formation could not be determined. It appears that they arise directly from the pinocytotic vesicles, which are generally formed at the peripheral ooplasm adjacent to the plasma membrane of the growing oocyte (Guraya, 1974). In the various electron microscopic studies on the oocytes of human and other primates, these vacuoles (or yolk vesicles) have been little studied for their presence, origin, structure, and nature (see Guraya, 1974). It appears that they are probably destroyed because of their watery contents, during the preparation of material for electron microscopy. However, some workers demonstrated the presence of several empty vesicles or vacuoles in the ooplasm of human oocytes, which do not show any material in their interior (Guraya, 1974). Some investigators have reported the presence of "protein crystals" or "crystalloid structures" in the oocytes of rhesus monkey and the human (Guraya, 1974). Recent studies on the oocytes of non human mammals have shown that ribosomes and RNA formed during oogenesis are apparently packed into lattices, fibrillar arrays, ladder-like structures or ribonuclear particles and remain masked for extended period of time (Guraya, 2000a). This may also be true for human vacuoles or yolk vesicles. The precise function of ooplasmic lattices or yolk vesicles is still controversial, as these are also believed to serve as yolk (Gosden and Bownes, 1995).

After the ooplasmic organelles have multiplied during oocyte growth, they are rearranged to form conspicuous gradients in the ooplasm of the fully mature oocyte (Figs. 9B and 9C) (Guraya, 1967b, 1970b, 1972, 1974), which were not described in the various electron microscope studies on oocyte development in human and other primates (Zamboni *et al.*, 1966). Similarly, they were not observed in ultrastructural studies on oocytes in other mammals (Guraya, 1985, 1997, 2000a). The animal half of the egg containing the nucleus (germinal vesicle) shows a greater concentration of yolk nucleus derivatives (or Golgi complexes), granular basophic substance (or granular endoplasmic reticulum and free ribosomes), mitochondria, some phospholipid bodies and spherical bodies (Fig. 9 B, C). Lipid bodies show some variations in number in the mature oocytes of different primates (Guraya, 1967b, 1970b, 1972, 1974, 1985, 2000a). Their number is relatively greater in the oocytes of rhesus monkey. In the mature oocytes of marmoset, the lipid bodies form a close anatomical association with the clear vacuoles (or yolk vesicles) (Guraya, 1967b, 1970b, 1974, 1985, 2000a). The physiological significance of this morphological relationship is required to be determined at the molecular level. The vegetal half of the egg is mostly occupied by fluid-filled vesicles. Guraya (1972) has suggested that gradients due to ooplasmic components (Figs. 9B and 9C) represent regional differences in the concentration and possibly the nature of informational molecules (RNA, DNA, and proteins), which may be of great physiological and developmental significance in embryogenesis, as they indicate local chemical and metabolic differences in the ooplasm to play some important roles in the regulation of gene expression during early embryogenesis. Comparative cellular and molecular studies are required to be made of gradients due to ooplasmic components in the fully-grown oocytes of young and aged women to determine their roles during early embryogenesis at the molecular levels.

2.2 NUCLEUS AND FORMATION AND STORAGE OF RIBONUCLEOPROTEINS

With the initiation of growth in the primordial oocyte, the nucleus (germinal vesicle) increases in size (Fig. 9A). However, the nuclear components, such as the large nucleolus, small nucleolus-like bodies, and lampbrush chromosomes, remain the same as those described for the nucleus of primordial oocyte (Chapter 1; Guraya, 1967b, 1970b, 1972a,b, 1974, 1985, 2000a; Fauser *et al.*, 1999). The growing oocyte shows a large spherical nucleus (germinal vesicle) with a prominent nucleolus and one or more smaller nucleoli (Figs. 9A, 9B, 9C and 11), and volume of the nucleus is increased in proportion of growth of oocyte. Throughout oocyte growth the nucleolus enlarges and shows progressive alterations in structure (Figs. 9A, 9B, 9C and 11), which are believed to be indicative of a period of intense ribosomal RNA synthesis. Actually conspicuous morphological and biochemical alterations occur in nuclear components such as the chromosomes, nucleolus and nuclear envelope during oocyte growth, which are related to the synthesis and accumulation of various RNAs (Guraya, 1972a,b, 1974, 1985, 1997, 1998a, 2000a; Fauser *et al.*, 1999). Growing oocytes are apparently more active having a prominent nucleolus (Figs. 9A, 9B, 9C and 11) and abundant ribonucleoproteins in the nucleus. The nucleolus shows fibrogranular network in small oocytes but becomes extensively fibrillar in nature in full growing oocytes, The large nucleolus develops vacuoles of different sizes in its central portion (Figs. 9A, 9B and 9C), which react negatively with all the histochemical techniques (Guraya, 1967b); basophilic material consisting of RNA and protein forms the cortex of a large nucleolus. In addition to the nucleoli, a fine granular substance consisting of RNA and protein is also seen in the nucleoplasm. The nucleolus vacuolization and the chromosome despiralization (to be described later here) during the intensive growth of oocyte are believed to indicate unraveling and increased transcription of rRNA as little transcriptional activity is seen in the fully-grown oocyte.

The chromosomes of growing oocytes, and also those in primordial oocytes (Chapter 1) remain in a more or less diffuse diplotene stage (Guraya, 1974, 1985, 2000a). The diplotene stage of nucleus comes to an end in response to the effect of LH secreted during the preovulatory period as will be discussed in (Chapter 3). Such a nucleus resembles an active interphase nucleus and supports the unique phase of oocyte growth and differentiation. The study of nucleus structure and transcriptional activity in relation to oocyte diameter in cattle has suggested that RNA synthesis appears to cease as the oocyte diameter is increased and consequently nucleoli are restructured from fibrogranular to dense fibrillar (Fair *et al.*, 1996, 1997). Such studies are required to be carried on the oocytes of young and aged women (Fauser *et al.*, 1999; Wolf and Zelinstski-Wooten.2001). But the observations with the light microscope show that growing oocytes in human and other primates possess larger chromosomes with well-developed lateral projections or loops (Guraya, 1974), suggesting that they also possess a lampbrush-type chromosomal configuration, which is similar to that found in lower vertebrates and invertebrates (Guraya, 1974, 1985, 1986, 1989).

The marked increase in the volume of the nucleus of oocytes in human and primates appears to be associated with the high metabolic activity of the lamp- brush chromosomes which actively incorporate labeled precursors of RNA and protein (Fauser *et al.*, 1999) as also described for other vertebrates (Guraya, 1974, 1985, 1986, 1989, 2000a). The dictyate stage typical of oocytes in rat and mouse (Guraya, 1974, 2000a; Gosden and Bowness, 1995) is believed to represent a highly diffuse lampbrush phase. The apparent absence of "cores" and their associated sheaths in rat can be interpreted as being due to the presence of very large, tortuous loops which could not be identified in thin sections (Guraya, 1974). These differences in the configuration of lampbrush chromosomes in different mammalian species simply reflect variations in the time at which the lateral loops form and the extent to which they develop (Guraya, 1974, 1985, 2000a). The observed differences in the radiosensitivity of oocytes among (1) species, and (2) among different stages of follicular development, are due to the degree of diffusion of the chromosome material in the form of lateral projections of the lampbrush type. Thus, oocytes in which the lampbrush loops are compact and possess a dense ribonucleoprotein sheath are resistant to radiation-induced cell death (human, monkey and guinea pig). Conversely, oocytes containing highly diffuse chromosomes with an attenuated ribonucleoprotein sheath are radiosensitive (e.g. mouse, rat). The precise way in which the mechanism operates is not yet known. It will also be rewarding to make a comparative cellular and molecular study on the lampbrush chromosomes in the growing oocytes of young and aged women (Fauser *et al.*, 1999; Bulletti *et al.*, 2001). The results of such studies will be very helpful to know the differences not only in their radiosensitivity as well as sensitivity to different culture media which are being extensively used for cryopreservations and culturing of human follicular oocytes, fertilized eggs these days (Wolf and Zelinstski-Wooten.2001). Such results will also be useful in the evaluation of transcription processes in their oocyte.

A large proportion of RNA and protein present in the growing oocytes of human and other primates and mammals (Guraya, 1972a,b, 1974, 1985, 1997a,b, 2000a; Yen *et al.*, 1999; Fauser *et al.*, 1999; Tevelde *et al.*, 2001) is produced by the lampbrush chromosomes and subsequently moves out of the nucleus into the ooplasm as supported by the correlation of results of ultrastructural, autoradiographic and biochemical including molecular studies. The biochemical and molecular studies have demonstrated the qualitative and quantitative aspects of synthesis and accumulation of stage-specific RNA species and proteins, histones, actin, and tubulin, putative intermediate filament protein, calmodulin, zona proteins, lactate dehydrogenase, creatine kinase, glucose- 6-phosphate dehydrogenase, lactate dehydrogenese and others during oogenesis of laboratory mammals (Fauser *et al.*, 1999; Guraya, 2000a). These various proteins, along with ribosomal proteins appear to form the stable, major bulk proteins of the oocytes. Comparative molecular and biochemical studies are required to be made of the problems of transcription and translation during oogenesis of young and aged women to define their roles during early embryogenesis (Bulletti *et al.*, 2001). But in laboratory mammals, major classes of RNA such as rRNA, tRNA, poly $(A)^+$ RNA, heterogeneous (hn) RNA and ribosomes show steady accumulation during oocyte growth but degradation increases once maturation initiates (Guraya, 2000a). Such studies are required to be made of major classes of RNA in the oocytes of young and aged women (Fauser *et al.*, 1999; Bulletti *et al.*, 2001).

rRNA forming approximately 63% of the total RNA observed in the ovulated mouse oocyte is synthesized in the nucleus at all stages of its growth (Gosden and Bownes, 1995; Fauser *et al.*, 1999; Guraya, 2000a). Similarly, ribosomal proteins constituting approximately 15% of the total proteins are synthesized at all stages of oocyte growth. Only 20% of the rRNA and ribosomal proteins produced in the growing oocyte occur in polysomal form and approximately 80% of them inactive in translation and are, therefore, accumulated for activation during early embryogenesis. The poly (A)-containing RNA in ovulated mouse eggs constitutes 6.6 and 8.3% of the total RNA (Gosden and Bownes, 1995; Guraya, 2000a); non-polyadenylated mRNA appears to form, a minor component. The large amount of mRNA accumulated in the mouse egg can support protein synthesis up to the 8-cell stage protein synthesis during this stage of development is known to be supported entirely by the various informational molecules formed and stored during oocyte growth. Although histochemical studies have demonstrated the accumulation of RNA and proteins (ribonucleoproteins) as result of activity of lampbrush chromosomes of nucleus during growth of human oocytes (Guraya, 1974), but comparable biochemical studies as made for rodents, are required to be carried out on the percentages of different RNA species during oogenesis of young and aged women and then their subsequent roles during early embryogenesis (Fauser *et al.*, 1999; Bulletti *et al.*, 2001).

Very divergent views exist about the site (s) and forms of accumulation of transitionally inactive rRNA and mRNA (Guraya, 1985, 1994, 2000a; Gosden and Bownes, 1995). Ribosomes and RNA formed during oogenesis are apparently packed into lattices, fibrillar arrays, ladder-like structures or ribonuclear particles and remain masked in the oocyte for extended period of time (Guraya, 1985, 2000a; Gosden and Bownes, 1995; Fauser *et al.*, 1999). The precise function of ooplasmic lattices is still controversial, as these are also believed to serve as yolk as already stated. Some workers believe that it is unlikely either that the lattices consist of ribosomes or that they have any significant amount of RNA. This suggestion is supported by the fact that yolk vacuoles, which are believed to represent ultrastructural lattices do not show RNA to be demonstrable with histochemical techniques (Guraya, 1974); they are filled with fluid which may be containing some proteins.

The formation of structural and other house-keeping proteins corresponds to the synthesis of a changing pattern of stage-specific proteins (Guraya, 1985, 1997a,b, 2000a; Gosden and Bownes, 1995; Fauser *et al.*, 1999). The growing oocyte is very active in the synthesis of proteins from nuclear and mitochondrial transcrips. Such studies are required to be made for the growing oocytes of young and aged women for knowing the differences if any to determine their role in early embryogenesis (Bulletti *et al.*, 2001). However, there is seen about 38-fold increase in absolute rate of protein synthesis in the growing oocyte of mouse (Guraya, 1997a,b, 2000a; Gosden and Bownes, 1995; Fauser *et al.*, 1999). Immunohistochemical analyses with the human DAZIA antiserum have shown that DAZIA protein is expressed at a cytoplasmic location in female germ cells (Nishi *et al.*, 1999). Avilable evidence suggests that the DAZIA gene is a participant in human oogenesis (Fauser *et al.*, 1999).

Detectable amounts of a number of enzymes involved in cellular house keeping functions are observed in the oocytes (Gosden and Bownes, 1995; Fauser *et al.*, 1999; Guraya, 2000a). While concentrations of some enzymes are low until embryogenesis (e.g., a-and b-glactosidase, b-glucuro-nidase, uridinekinase, hypoxanthine guanine phosphoribosyltransferase (HPRI), adenine phosphori bosyltransferase, others occur in greater amounts in mouse eggs and remain stable or increase more after fertilization (e.g., glucose-6-phosphate dehydrogenase, glucose phosphate isomerase (GPI), phosphoglycerate kinase, creatine kinases, lactate dehydrogenase (LDH). LDH is present in the fully-grown oocyte of mouse and is responsible for 2-5% of total protein synthesis at some stages. Such studies are required to be carried out on the proteins of oocytes in young and old women (Bulletti *et al.*, 2001) as for diagnosing genetic defects by enzyme assay of embryonic biopsies, it is important to determine the amount of the maternally encoded enzymes that remains as compared to that being expressed by the human embryo, which expresses its genome after the 4-to-8-cell stage (Braude *et al.*, 1988, Chapter 6). Significant amounts of maternally derived enzymes such as HPRT and GPI may be found as late as the 8-cell stage in the mouse. But the comparative aspects of enzymes in the eggs of young and old women are required to be investigated (Bulletti *et al.*, 2001). But there is six times as much GPI in human, compared with mouse oocytes whereas the difference in cell volume is only three fold. With exception of action, our knowledge is meagre about the precise function of maternal transcripts in the synthesis of structural proteins in human embryos of young and aged women (Chapter 6). The results of such comparative studies will be very rewarding to reveal the molecular basis of embryonic disorders as a result of aging of human primordial oocytes (Guraya, 1999a).

Some of the proteins found in the oocyte may not be synthesized there but taken up from the FF as in developing oocytes a macromolecular component can be endocytosed by specific mechanisms from the surrounding body fluids or blood. Despite the ultrastructural and cytological evidence of endocytosis in mammalian oocytes, there are few clues about the identity of molecules being transported (Guraya, 1998b; Fauser *et al.*, 1999). It is still required to be determined whether the storage proteins produced in other organs such as liver are transported to the oocyte for storage. Therefore, it will be important to look first-at the more abundant proteins in the human oocytes with a possible storage function such as the protein lattices of rodent oocytes (Gosden and Bownes, 1995; Guraya, 2000a). Ostorlund and Fried (2000) have demonstrated the presence of transforming growth factor (TGF) beta receptors types I and II and the substrate proteins smud 2 and 3 are present in human oocytes, supporting their hypothesis that the TGF beta I in FF may interact with the oocyte and preimplantation embryo via TGF beta receptors and that TGF beta signaling may be important for the development of the oocyte and the preimplantation embryo.

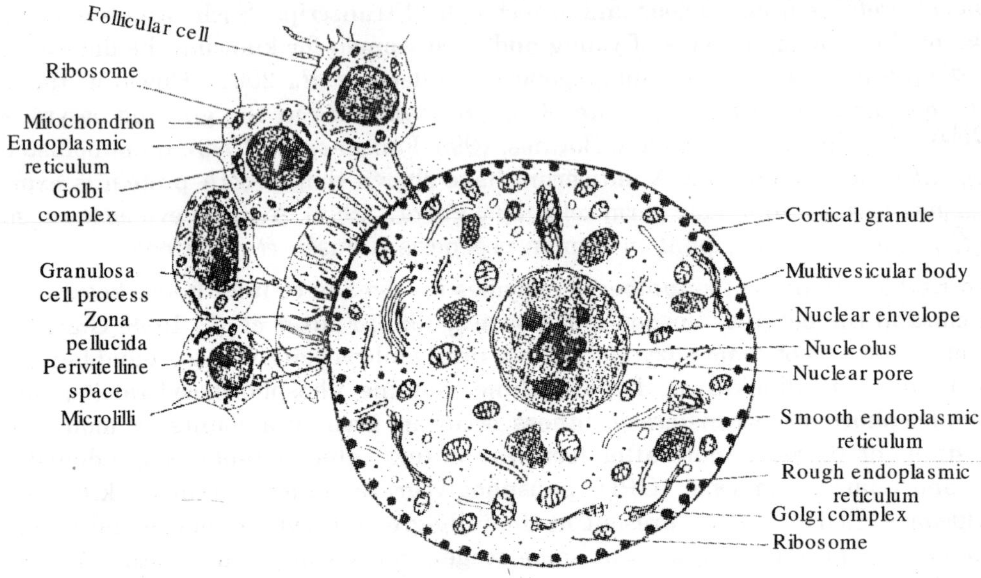

Fig. 11

Some early proteins whose function becomes critical in later embryonic development include histone H_4, some of the ribosomal protein and a so-called germinal vesicle associated protein (GVAP), which are sequestered in the nucleus and used during maturation and development (Fauser *et al.*, 1999; Guraya, 2000a). Some of the proteins are required for the differentiation of the oocyte itself, some for interacting with the surrounding granulosa cells and yet others will be needed for early development (Chapter 6); many are significant for growth in a range of cell types, although some which are specific to the oocyte have been studied. There is also seen the early synthesis of three glycoproteins which form the zona pellucida and function during fertilization. Rapid advances in the study of genes encoding these proteins have contributed to a better understanding of ZP proteins as will be discussed in Section 2.3. There is a need of study of genes encoding various proteins in the growing oocytes of young and aged women (Fauser *et al.*, 1999; Bulletti *et al.*, 2001). Such a knowledge of protein content is required for the better interpretation of *in vitro* culture data, metabolic and protein synthesis studies as well as for the assessment of embryo growth, infant disorders and viability in human (Wolf and Zelinski-Wooten, 2001). The mouse growing oocyte synthesizes plasminogen activators (tPA), which are apparently secreted around fertilization to play some specific roles (Gosden and Bownes, 1995). The enzyme induction results from the translational activation of a preformed tPA, mRNA during oogenesis. The identification of tPA, mRNA represents the first characterization of a dormant maternal mRNA in mammalian oocytes. The rise in nucleoplasmic Ca^{2+} concentration occurs as a result of the influx Ca^{2+} that is released in the perinuclear cytoplasm, not Ca^{2+} release from nuclear envelope to the nucleoplasm (Guraya, 2000a). The transfer of various informational molecules from the nucleus into the ooplasm may be facilitated by the simultaneous development of several pores in the nuclear envelope of growing oocytes studied with the electron microscope (Guraya, 1974, 1985, 2000a). The RNAs and proteins of nuclear origin, which accumulate in the ooplasm of growing oocytes, are known to play significant roles during early embryogenesis (Chapter 6). The RNAs produced by the oocyte nucleus during oogenesis act as inducers as will be discussed in Chapter 6.

2.3 ZONA PELLUCIDA

One of the most conspicuous changes occurring during human oogenesis is the secretion of glycoprotein coat, zona pellucida (ZP), which forms a general feature among mammals (Figs. 10, 11, 12 and 13) (Guraya, 1974, 1985, 1997a,b, 2000a; Dietl, 1989; Fauser *et al.*, 1999; Van den Hurk *et al.*, 1999; Tevelde *et al.*, 2000). The ZP first appears in human oocytes that have a cuboidal layer of granulosa cells (Guraya, 1967b, 1969a). It starts its development as islands of fibrillar material situated in spaces between adjacent granulosa cells and the oocyte surface (Guraya, 1974, 1985, 1997, 2000a; Dietl *et al.*, 1989). The formation of ZP corresponds closely to the growth of the human oocyte (Figs. 12A and 12B). The granulosa cells or oocyte or both have been suggested to be involved in the formation of ZP (Guraya, 1974; Dietl, 1989; Sharma and Guraya, 1995) as will also be discussed later on. Progressive enlargement as well as confluence of lacunae or islands leads to the formation of a continuous ZP which consists of a carbohydrate-protein complex (Figs. 12A and 12B) (Guraya, 1967b, 1969a, 1974, 1985, 1997, 2000a; Dietl, 1989; Fauser *et al.*, 1999); some RNA is also present which may be due to the presence of some ribosomes in the processes of granulosa cells traversing the ZP (Fig. 11).

The materials of the ZP in the rhesus follicle show a bipartite structure (Guraya, 1974). The material adjacent to the oocyte has a homogeneous structure, whereas the material adjacent to the granulosa cells contains, in addition, a granular, electron-dense material. In the human, the layer adjacent to the granulosa cells is more flocculent than that next to the oocyte (Guraya, 1974). It has also been shown that the outer layer contains acid mucopolysaccharides, while the inner layer contains neutral mucopolysaccharides. It has been suggested that the granulosa cells secrete the material of the ZP but since frequently there are gaps between the granulosa cells and the presence of this material has suggested that the oocyte plays a part in polymerizing it. PAS-positive material is found in the granulosa cells and also morphologically similar material in the ZP (Guraya, 1974; Sharma and Guraya, 1995) and consequently it has been suggested that this material is secreted by the granulosa cells. However, since the outer layer contains acid mucopolysaccharides and the inner layer, neutral mucopolysaccharides, the oocyte is also involved as supported by the peripherally arranged Golgi bodies of the oocyte (Figs. 10 and 11), although the granulosa cells show well-developed granular endoplasmic reticulum (Guraya, 1974; Sharma and Guraya, 1995). A careful examination of granulosa cell cytoplasm, especially of the processes, reveals the presence of some flocculent and fibrillar material which is apparently similar to that of a ZP (Guraya, *et al.*, 1974; Sharma and Guraya, 1995). Corresponding to the deposition of ZP material, the granulosa cells develop ultrastructural features of their cytoplasmic organelles and nuclear components (Figs. 10 and 11) which can be related to high metabolic activity, especially to protein synthesis. As the ZP widens and becomes more continuous, granulosa cell processes containing some electron dense material project into it obliquely (Fig. 11). These cytoplasmic extensions also spread between and into adjacent granulosa cells (Guraya, 1974), some of them are also completely incorporated into adjoing granulosa cell, suggesting transfer of cytoplasm from one granulosa cell into another.

The zona material is now well known to be mainly formed by the activity of Golgi complexes and granular endoplasmic reticulum of the growing oocyte (Guraya, 1974, 1984, 1985, 1997, 2000a; Dietl, 1989; Gosden and Bownes, 1995). But the contribution of granulosa cells to the formation of ZP still remains to be determined more precisely and will be discussed later on. Finally ZP becomes a denser and thicker network of interconnected filaments completely surrounding the oocyte and largely separating it from the granulosa cells. The oolemma and surrounding cumulus cells form microvilli and processes respectively, which traverse the ZP (Fig. 11) as will be discussed in Section (2-4). But the contact between the oocyte and the corona radiate cells is maintained via functional complexes which are involved in the communication between oocyte and surrounding cumulus cells as small molecules involved in oocyte metabolism and regulation of meiosis pass from granulosa cells into oocyte (Guraya, 1985, 2000a) as will be discussed in Section 2.4.

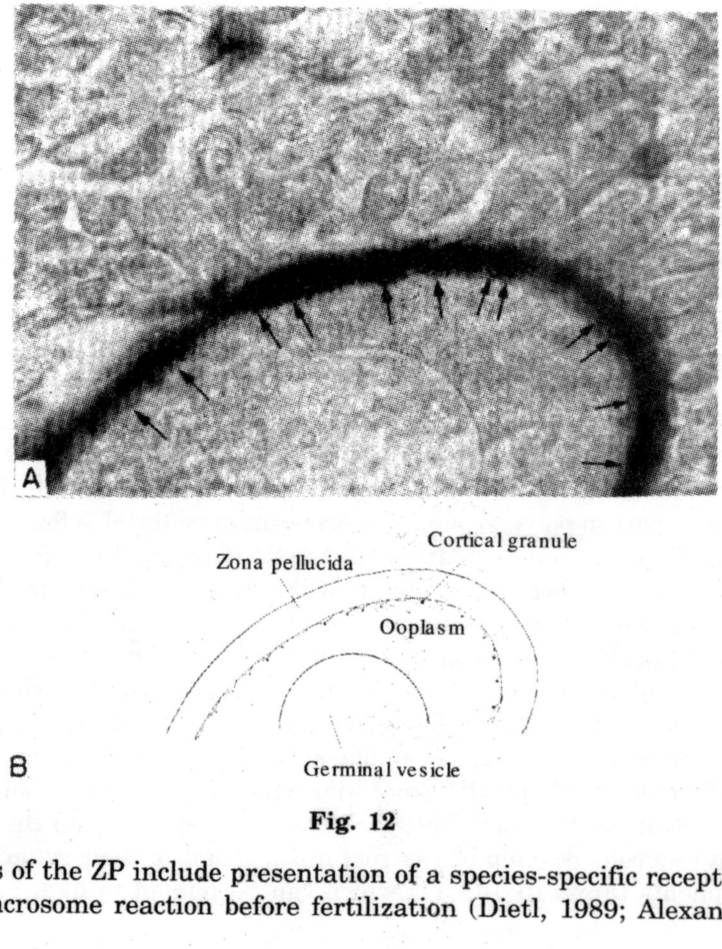

Fig. 12

The functions of the ZP include presentation of a species-specific receptors to sperm and induction of the acrosome reaction before fertilization (Dietl, 1989; Alexander *et al.*, 1989;

Fauser *et al.,*1999; Tevelde *et al.*, 2000 Guraya, 2000a, b; Chapter 5). Once the egg is fertilized, the zona becomes more harder, more insoluble and possibly less permeable as a result of steric alterations in glycoproteins, which are of great courent interest for their synthesis, chemistry and functions. Their composition is quite variable as a result of carbohydrate moieties added to a small number of polypeptides. The proteins of ZP have been isolated and chemically characterized. Differences are reported in composition of oligosaccharides, amino acids, glycopeptide maps, biosynthetic precussors, immunological attributes and biological functions of ZP glycoproteins (Alexander *et al.*, 1989; Fauser *et al.,*1999; Gupta *et al.*, 2000). The glycoproteins of the ZP form the major class of proteins glycosylated secreted during oocyte growth as already described (Dietel, 1989; Gosden and Bowness, 1995; Fauser *et al.,*1999; Guraya, 2000a). The glycoproteins moieties are generally designated ZP_1, ZP_2, ZP_3 and ZP_4 with different molecular weights depending upon the species (Dietl, 1989; Alexander *et al.*, 1989b; Gosden and Bownes, 1995; Fauser *et al.,*1999; Gupta *et al.*, 1999; Bousken *et al.*, 1999; Guraya, 2000a). They are not only different molecular species but also have different biological functions and variable patterns of glycosylation. The ZP glycoproteins play a crucial role in the initial attachment followed by tight binding of spermatozoa to the oocyte and subsequent cascade of events during fertilization (Chapter 5) (Dietl, 1989; Alexander *et al.*, 1989; Gupta *et al.*, 1999; Fauser *et al.,*1999). Due to their crucial involvement during fertilization, ZP glycoproteins have been proposed as candidate antigens to develop antibodies and hence an immunocontraceptive; vaccine for fertility regulation (Gupta *et al.*, 1996, 1997a,b, 1999, 2000; Paterson *et al.*, 2000).

One of the glycoproteins designated ZP_3 functions both sperm receptor and acrosome reaction during fertilization of mammalian ovulated eggs (Guraya, 2000b; Dietl, 1989; Alexander *et al.*, 1989; Fauser *et al.,*1999). The cDNA has been used to elucidate the genomic organization of this specific zona protein gene (reviewed by Sidhu and Guraya, 1991). The ZPs are obviously representatives of a family of highly conserved genes as 60% of the amino acid sequence of human murine ZP_3 is similar. Studies have been made of molecular cloning and expression of genes encoding ZP glycoproteins especially ZP_2 and ZP_3 in different species of mammals and human (Dietl, 1989; Dean, 1993; Afzalpurkar and Gupta, 1997; Afzalpurkar *et al.*, 1997; Kaul *et al.*, 1996; Kaul *et al.*, 1997; Jethanandari *et al.*, 1998; Gupta *et al.*, 1999; Fauser *et al.,*1999; Gupat 1999; Tevelde *et al.*, 2000). The genes encoding ZP_2 and ZP_3 are conserved among mammals. Taking cross hybridization of nucleic acid sequences as criteria the degree of conservation of ZP_3 is variable with pig, dog, cow and human genes. The sequence of amino acids of ZP_2 and ZP_3 show species variations (Dean, 1993). The genes for ZP glycoproteins have been cloned and sequenced from several species. Human ZP_1 (h ZP_1) gene spans 13kb having 19 exons and is transcribed into a 2235 nucleotides (nt) mRNA encoding a polypeptide of 745 amino acid (aa) residues (Liang and Dean, 1993; Fauser et al., 1999). Human ZP_2 (h ZP_2) gene spans 11 kb and is composed of 12 exons. The h ZP_2 transcript has an open reading frame (ORF) of 1600 ht and encodes a polypeptide 54 aa residues. Human ZP_3 (h ZP_3) gene spans 18.3 kb containing 8 exons encoding a polypeptide of 424 aa residues (Gupta *eta al.* 1999; Gupta 1999). ZP_3 is present in oocytes at all stages of folliculogenesis as well as in granulosa cells as determined by the expression and localization of glycoproteins and mRNA in cynomolgous monkeys (*Macaca fascicularis*). Presence of ZP_3 on primordial follicles and granulosa cells of the human ovary has also been demonstrated by using specific polyclonal as well as monoclonal antibodies, supporting their role in the formation of ZP (Gupta *et al.* 1999; Gupta 1999; Fauser *et al.* 1999; Tevelde *et al.* 2000. These observations are at variance with the one from mouse model (as already discussed) and thus require further clarification on the possibility of expression of ZP glycoproteins by other ovarian associated cells such as granulosa cells in addition to the oocyte. The precise regulation of expression of three zona transcripts is still required to be determined. But the concurrent accumulation suggests that common transcriptional regulatory events may be involved in the regulation of gene expression (Gupta *et al.*, 2000).

In order to elucidate further the potential of ZP glycoproteins as candidate antigens, for designing of an immunocontraceptive vaccine, (Alexander *et al.*, 1989). Gupta *et al.*, (1999) have discussed the characterization of the bonet monkey (*Macaca radiata*) bm ZP_1, bm ZP_2, and bm ZP_3. bm ZP_1, bm ZP_2, and bm ZP_3 possess 92.0%, 92.2%, and 93.9% amino acid sequence identity with human (h) h ZP_1, h ZP_2, and h ZP_3, respectively. The major difference between bm ZP_1 as compared to h ZP_1 is the deletion of the 28 amino acid domain corresponding to the 100-127 amino acid residues. Comparison of ZP_3 deduced amino acid sequence from various species has shown that domain corresponding to 318-352 amino acid residues is least conserved. But this domain is well conserved between bonnet monkey and human ZP_3. bm ZP_1, bm ZP_2 and bm ZP_3 are expressed as polyhistidine fusion proteins excluding sequences coding for the N-terminal signal sequence and C-terminal transmembrane-like domain and T_5 promoter using p QE-30 vector in *Escherichia coli*. The antibodies generated against *E. coli.* expressed recombinant ZP_1, ZP_2 and ZP_3 recognized bonnet monkey zonae (Gupta *et al.*, 1999, 2000). Peterson *et al.* (1998) have made evaluation of contraceptive potential of recombinant human ZP_3 and human ZP_3 peptides in a primate model in relation to their safety and efficacy.

In the lines as discussed above, it will be rewarding to make comparative molecular, biochemical and immunological studies of zona glycoproteins (ZP_1, ZP_2 and ZP_3) in the ovaries of young and aged women. The results of such studies will help us to know the effects of aging of primordial oocytes (as discussed in Chapter 1) on the changes in the chemical composition and immunology of these glycoproteins which are now well known to be synthesized mainly by the growing oocytes as already discussed.

2.4 FUNCTIONAL INTERRELATIONSHIPS OR METABOLIC COOPERATIVITY BETWEEN GROWING OOCYTE AND GRANULOSA CELLS

The highly active metabolic state of the growth of the oocyte is believed to impose demands on its plasma membrane, which are far in excess of the limited uptake features of this membrane (Guraya, 1985, 1997, 1998a, 2000a; Fauser *et al.*, 1999; Yen *et al.*, 1999). The oocytes isolated from direct contact with surrounding granulosa cells do not grow normally *in vitro* (Wolf and Zelinski-Wooten, 2001) supporting the influence of granulosa cells on oocyte growth. Various electron microscopic studies of the precise nature of heterogeneous contacts between the oocyte and granulosa cells have shown that numerous microvilli and micropapillae on the oocyte surface extend across the ZP and come in close contact with the processes of granulosa cells (Figs. 11, 13 (A-D)) and small gap junctions are developed at points of juxtaposition of the oolemma and granulosa cells (Fig. 11). Actually, the oolemma of small oocytes is fairly smooth but becomes increasingly folded during growth by forming uniform cover of microvilli with cores of microfilaments from junctional contact with corona cell processes traversing in the opposite direction. The corona cells are connected with the oocyte by one or few cytoplasmic processes across the ZP, which terminate in button a bulbous (Figs. 13 (A-D)) or indent it deeply (Figs. 11, 13 (A-D)) (Gosden and Bowness, 1995; Dietl, 1989; Sharma and Sawhney, 1999; Guraya, 2000a). These button like swellings containing few organelles terminate on the cell surface or in pits in the oolemma buttons either attach to oocyte plasmalemma superficially (Figs. 13 (A-D)) or indent it deeply (Dietl, 1989; Gosden and Bownes, 1995; Guraya, 2000a). Along the apposing membranes junctional complexes are formed. Such studies on the morphological relationships between corona cell processes and oocyte plasmalemma, which have been made mostly for rodents and ruminants, are required to be made for young and aged women (Sharma and Guraya, 1990; Fauser *et al.*, 1999; Sharma and Sawhney, 1999) as the subcellular specialization of juxtaposed plasma membranes of the oocyte and adjacently placed granulosa cells provides the means of communication between them as the oocyte does not have direct cytoplasmic bridge to the somatic supporting cells (or corona radiata cells). It is called a gap junction. These morphological specializations of the plasma membrane of the oocyte and its surrounding corona cells constituting the coupling pathways, which still remain to be investigated not only for young and aged women (Tevelde *et al.*, 2000; Bulletti *et al.*, 2001) but also for more species of mammals (Guraya, 2000a), are believed to facilitate introduction of extracellular substances such as nucleotides, sugar, fatty acids, amino acids etc. required for various metabolic activities of growing oocyte (Guraya, 1985, 1997a,b, 2000a; Gosden and Bownes, 1995). Thus, the porous ZP allows exchange of molecules with follicular fluid and foot processes of cells in the corona radiata of an acicular follicle, the foot processes terminate as buttons on the oolemma with adhesive (intermediate)-type junctions as well as gap junctions. The gap junctions formed between the granulosa cells provide low resistance electrical pathways and pores to permit exchange of ions and molecules. Thus, intercellular passage of molecules through the junctions provides the high substrate uptake required by the oocyte during its growth phase. This is also called metabolic cooperativity, which occurs when a metabolite is transported from one cell to another, via gap junctions, to take part in a metabolic activity (Eppig, 1994; Guraya, 2000a). Actually the granulosa-oocyte complex is a metabolically coupled unit in which ions, amino acids, nucleotides etc. are transported between the compartments without affecting the distinctive macromolecular phenotype of other cell type. The decreased ability of naked oocytes to take up amino acids directly from the medium *in vitro* is the result of a deficiency of the energy-dependent A-transport system that is normally compensated for by supplies from the granulosa cells via gap junctions. Gap junctions may also provide transport mechanism by which one cell can regulate the function of another cell type (see Guraya, 1999b; Fauser *et al.*, 1999). For example, cAMP is the molecule that is transmitted from the granulosa cells via the heterogeneous gap junctions affects the activity of the oocyte (Guraya, 1985, 1997a, 2000a; Eppig, 1994). The observations made by Grieshaber *et al.* (2000) indicate that the adenylyl cyclase/cAMP signal is necessary and sufficient for FSH-dependent granulosa cell differentiation, including massive reorganization of the actin cytoskeleton and changes in the cell morphology and cell-to-cell interactions. There is no evidence that the phospholipase C signal and Ca^{2+} mobilization are involved in this process. FSH induces the formation of lamellipodia and filopodia via the adenylyl cyclase/cyclic adenosine monophosphate signal. Selective uptake of amino acids, cAMP and various other chemical compounds into oocytes from cumulus cells as described for some mammalian species (Guraya, 1985, 1997a, 2000a; Eppig, 1994; Gosden and Bownes, 1995) are required to be investigated for the oocytes of young and aged women for determining their precise roles in those two situations (Tevelde *et al.*, 2000; Bulletti *et al.*, 2001). The results of such studies will help us to know the effects of aging on the various metabolic activities of their growing oocytes, which are not known (Fauser *et al.*, 1999)

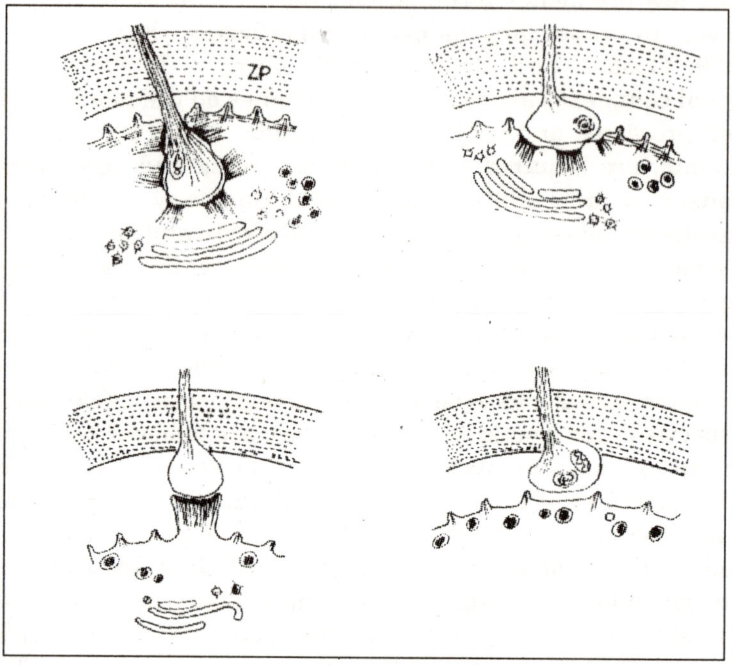

Fig. 13

The proliferation morphologenesis and differentiation of the cumulus cells are now known to be well influenced by local factors released by the oocyte (Bachvarova *et al.*, 1993; Dietl, 1989). The major goal for future investigations must be to elucidate the comparative nature of signalling molecules passing between oocytes and granulosa cells (Fauser *et al.*, 1999), which regulate the development of the follicle as a unit in the ovaries of young and aged women (Tevelde *et al.*, 2000; Bulletti *et al.*, 2001). The extent of metabolic cooperativity is increased during oocyte growth *in vivo* and *in vitro* with the extent of intercellular communication (gap junction formation) as already discussed. But this metabolic cooperativity or union between the corona cells and oocyte is broken within a few hours of the ovulatory gonadotrophin surge when the corona cell processes are withdrawn and the vitellus shrinks (Guraya, 1985, 1997, 2000a; Dietl, 1989).

The surface properties of the oolemma are also of great importance for further development and one of the most crucial the ability to bind the fertilizing spermatozoa (Dietl, 1989; Sidhu and Guraya, 1991; Fauser *et al.*, 1999; Guraya, 2000b; Wolf and Zelinski-Wooten, 2001). Therefore, the study of chemical properties of oolemma during oogenesis is of great current interest as supported by the evidence that peptides with affinity for fibronectin-like molecules bind to the oolemma, suggesting that the ligand on this membrane is an integrin (Fauser *et al.*, 1999). The acquisition of fusibility by the rat oolemma occurs during the growth period involving the appearance of DE (a glycoprotein, 37 kDa) binding components on the oocyte surface (Cohen *et al.*, 1996). This study provides novel information on the molecular mechanism by which the mammalian egg plasma membrane becomes competent to fuse with sperm during oogenesis. Cecconi *et al.* (1996) have shown that in mice, the maturation of both oolemma and plasma, required for successful fertilization is acquired together with the development of antral follicles and appears to be affected by cumulus cells during meiotic prophase. The protein pattern in the human plasma membrane appears qualitatively limited to 13 species and quantitatively their amount decreases during oocyte preovulatory maturation (Ji *et al.*, 1997). A great polymorphism from one oocyte to another is observed. The protein pattern is highly conserved between human, hamster and mouse oocytes. But the functional significance of this protein pattern of oolemma is required to be determined in the oocytes of young and aged women (Tevelde *et al.*, 2000; Bulletti *et al.*, 2001) as the attention to the molecular aspects of oogenesis is becoming increasingly successful (Dietl, 1989; Gosden and Bownes, 1996; Fauser *et al.*, 1999; Guraya, 2000a).

The contribution of granulosa cells in oocyte growth has been a matter of some controversy. Correlated morphological and histochemical studies have shown that during follicle growth the granulosa cells fulfil the criteria for cells involved in protein synthesis (Guraya, 1971a, 1974, 1997, 1998a, b, 2000a). Part of this synthesis activity may be concerned with the formation of ZP as discussed in Section 2.3. Some of these substances synthesized by the granulosa cells may also enter the oocyte by some mechanism as well be discussed later; although the egg is capable of making some of the informational molecules itself (Section 2.2). Ultrastructural charges, which occur in the oocyte surface and its adjoining granulosa cells during oocyte growth, support this idea (Guraya, 1974, 1985, 1997, 1998a, 2000a). The long processes originating from the granulosa cells traverse the ZP to the surface of the oocyte, and in some cases penetrate deeply into the ooplasm to terminate a few microns from the oocyte nucleus (Figs. 11 and 12) (Guraya, 1974, 1985, 1997, 1998a, 2000a; Sharma and Sawhney, 1999). It has been suggested that the microvilli and granulosa cell processes have some function in the incorporation of nutritive material into the egg (Guraya, 1974, 1985, 1997, 2000a). The granulosa cell processes interdigitate with microvilli produced by the egg surface. Corresponding to these ultrastructural alterations in the oocyte surface and its adjoining granulosa cells, there are also developed activities of phosphatases at these sites, which are believed to be involved in the transfer of substances across the cellular membranes (Guraya, 1974). These enzyme activities presumably associated with microvilli and granulosa begin to decrease or disappear after the oocyte has attained its full growth, supporting their role in the transport of substances from the granulosa cells into the oocyte.

Vesicular profiles seen within the ooplasm contain electron opaque, granular material which is similar to that in the granulosa cell processes. This granular material may be composed of RNA and protein, which have been demonstrated histochemically in the ZP (Guraya, 1967b, 1970b, 1974). This suggests the possibility of infiltration of some proteins and informational materials (ribonucleo proteins) into the oocyte by diffusion, active transport or pinocytosis. The maternal proteins and other substances contribute to the growth of the oocyte by a process in which the ends of granulosa cell projections are incorporated into the egg (Guraya, 1974, 1985, 1997a,b, 1998b, 2000a). This would involve pinocytosis or phagocytosis, since there is no direct cytoplasmic continuity between the cell processes and the oocyte.

Small granules from the vicinity of the Golgi material of the corona radiata cells have been described by several workers (Guraya, 1974, 1985, 2000a). These granules have been interpreted by different investigators as being Golgi bodies, ribonucleoproteins, mitochondria or fat droplets (Guraya, 1974, 1985). Guraya (1970d) studying histochemically the transport of lipids into the developing oocyte of human and other primates, found that the infiltrating granules are lipid bodies composed of phospholipids as also reported for other mammals (Guraya, 1974, 1985, 1997a,b, 1998a, 2000a). The other workers did not observe lipid bodies passing through the zona pellucide into the oocytes of primates and other mammals. Infiltrating lipid bodies were apparently not fixed in osmium tetroxide, which was used in electron microscope studies (Guraya, 1974, 1985, 2000a; Sharma and Sawhney, 1999). Since the infiltration of lipid granules corresponds to the rapid growth of the oocyte they apparently supply phospholipids needed for the construction of membranes of ooplasmic organelles and other developmental processes as suggested by Guraya (1974, 1985, 1997, 2000a). The rapid inflow of lipids through the zona pellucida into the oocyte of cat and dog is closely related to the deposition of lipid yolk (Guraya, 2000a).

The visible functions of corona cells appear to be (1) to secrete lipids, proteins, glycogen and nucleic acid molecules, and (2) then to facilitate their migration along the corona cell processes into the oocyte (Guraya, 1974, 1985, 1997a,b, 1998b, 2000a; Fauser *et al.*, 1999). It has not been determined how incoming lipids and other chemical substances manage to pass through the plasma membranes of both the processes of corona cells and oocytes, as no direct continuity between them has been seen with electron microscopy (Guraya, 1974, 1985, 1997a,b, 1998b, 2000a; Gosden and Bownes, 1995). However, some suggestions in regard to the mechanism of passage of nutrient materials into the developing oocyte have been made (Guraya, 1974, 1985, 2000a).

The cells forming the cumulus under the normal preovulatory stimuli undergo a change, which alters their ability to grow in tissue culture (Guraya, 1974, 1985, 2000a, Sharma, 2003). The nature of this change is not known and further studies are required to define the problem as the cells of the cumulus apparently have a very definite life span after they come under the influence of specific preovulatory hormones. During this period, the number of pycnotic nuclei increases. The oocyte has been shown to produce soluble factors, which regulate a number of processes in follicular development, including cumulus expansion in the preovulatory period (Elvin *et al.*, 2000). These authors have discussed the similarities and differences in sequences, expression and function of the oocyte expressed TGF beta family members with respect to regulating folliculogenesis. The TGF-alpha and EGF receptor are expressed in the primordial to antral follicles, indicating a role of TGF-alpha in regulating follicular development through binding to the EGF receptor (Qu *et al.*, 2000a). Freeze thawing does not substantially charge immunoreactivities for TGF-alpha, EGF and EGF receptor in frozen ovarian tissue (Qu *et al.*, 2000b). These authors have not observed any significant difference in the immunohistochemical staining for EGF receptor in ovarian tissue before and after cryopreservations. The results obtained by Aaaitoonen *et al.* (1999) are consistent with the suggestion that growth differentiation factor-9 (GDF-9) and GDF-9B may regulate human folliculogenesis in a manner specific to the ovary. Garridi *et al.* (2000) have demonstrated that the follicular environment is different in cases with endometriosis and suggested that infertility in patients with endometriosis may be related to changes within the oocyte, which in turn, result in embryos of lower quality, and with a reduced ability to implant.

3

Oocyte Maturation

The molecular and cell biology of the oocyte growth in human is discussed in Chapter 2. During growth, oocytes synthesize and accumulate a variety of transcripts (RNAs) and proteins that may be required for general housekeeping purposes, important for oocyte growth or some aspect of function, or accumulation for use during embryogenesis (Wassarman and Albertini, 1994, Heikinheimo and Gibbons, 1998; Fauser *et al.*, 1999; Guraya, 2000a; Wolf and Zelinski-Wooten, 2001). In other words, the oocyte developmental competence is acquired both during follicle development before meiotic resumption, and during meiotic progression, concurrent with nuclear maturation (Schramm and Bavister, 1999). This chapter discusses the molecular aspects and signals resulting in prepairing the oocyte with the ability to complete maturation of the nucleus and cytoplasm. Arrest and reinitiation of oocyte maturation are related closely to the progress of folliculogenesis and ovulation (Guraya, 2000a). Somatic cells in the ovarian follicle perform important and variable functions in the regulation of these events (Eppig *et al.*, 1996; Guraya, 1997a,b, 1998a). Whitaker (1996) has discussed the regulation of meiotic arrest, which can be usefully formulated in terms of the interaction between cell signaling mechanisms and the protein machinery that regulates the cell cycle (Fauser *et al.*, 1999). Much of what we know about cell messengers, specially calcium and the cell cycle proteins comes from the studies on oocytes. Maturation prepares the oocyte with the ability not only to complete meiosis but also undergo fertilization, and prepares male and female haploid chromatin for syngamy and embryo development (Heikinheimo and Gibbons, 1998; Fauser *et al.*, 1999; van den Hurk *et al.*, 1999; Guraya, 2000a; Wolf and Zelinski-Wooten, 2001). The structural and regulatory proteins required to complete these processes are largely the result of transcription during an earlier stage of oocyte growth (Chapter 2) with translation or post-translational changes occurring immediately prior to a developmental process in response to a specific signal. Other essential proteins especially those with structural or housekeeping functions may accumulate for a long period in the oocyte (Chapter 2). Thus, the cytoplasm of the oocyte plays an important role in oocyte maturation as various ooplasmic factors play significant roles in the regulation of events of fertilization, the transition from meiosis to mitosis and early embryo development as will be discussed in chapters 5 and 6. Failure of oocyte maturation (or cytoplasmic maturation defects) can, therefore, cause anomalies such as absence of sperm chromatin decondensation, failure of male pronuclear formation, polyspermy, and inability of embryo to complete development (Fauser *et al.*, 1999; Guraya, 2000a; Wolf and Zelinski-Wooten, 2001).

The oocyte maturation *in vivo* and *in vitro* occurs in close relation to the growth and differentiation of the follicular unit as the growing oocytes of mammalian ovary develop the ability to undergo meiosis or maturation around the time of antrum formation under the influence of FSH (Guraya, 1997a, 2000a; Fauser *et al.*, 1999; van den Hurk *et al.* 1999; Tevelde *et al.*, 2000). Wolf and Zelinski-Wooten, (2001) have shown the progressive acquisition of meiotic competence in the equine oocyte during antral follicle growth. This is called "meiotic competence", which is not only age related but also related to follicular growth and is promoted by oestrogen. Such studies are required to be carried out on the comparative aspect of acquisition of meiosis competence in the oocytes of young and aged women as the primordial oocytes undergo aging changes (Guraya, 1999a, 2000a; Fauser *et al.*, 1999; Tevelde *et al.*, 2000; Bulletti *et al.*, 2001, Chapter 1). However, according to Volarcik *et al.* (1998), the human oocyte appears to be particularly prone to meiotic errors and the incidence of these errors is strongly influenced by maternal age, as they have observed aged-related effects on the meiotic process in oocytes from unselected follicles. Specifically, in oocytes obtained from donors over the age of 35 years, when the primordial oocytes show conspicuous aging changes as discussed in Chapter 1. The majority of the oocytes extruded a first polar body in culture and arrested at second meiotic metaphase. Aberrations in spindle formation and chromosome segregation occur at the first meiotic division, which are influenced by donor age *in vitro* stored oocytes (Wolf and Zelinski-Wooten, 2001) as is in oocytes undergoing meiotic maturation *in vivo*. An age-related decline in the process of folliculogenesis results in reduced oocyte quality and that the well characterized age-related increase in meiotic non-disjunction is one symptom of compromised oocyte growth and quality for normal development (Tevelde *et al.*, 2000; Bulletti *et al.*, 2001). Schramm and Bavister (1999) have demonstrated that oocyte development competence is likely acquired both during follicle development, before meiotic maturation, and during meiotic progression, concurrent with nuclear maturation.

The meiotic competence of marmoset monkey oocytes is related to follicular size and oocyte-somatic cell associations (Gilchrist *et al.* 1995). Follicular size and quality influence the quality of follicular fluid to support cytoplasmic maturation in bovine oocytes (Guraya, 2000a). The molecular mechanism by which the oocyte within the preovulatory follicle develops meiotic competence is of great current interest (Heikinheimo and Gibbons, 1998; Fauser *et al.*, 1999). Lack of adequate knowledge on this aspect restricts our ability to obtain normal maturation of human oocytes under controlled *in vitro* conditions (Wolf and Zelinski-Wooten, 2001). Incomplete or otherwise abnormal oocyte maturation may cause embryonic losses during pregnancy, and may also result in failure of pregnancy when embryos produced by *in vitro* fertilization are transferred (Wolf and Zelinski-Wooten, 2001). This is also true for the human as the primordial oocytes undergo aging changes which may lead not only to embryonic losses but also to embryonic abnormalities (Guraya, 1999a; Tevelde *et al.*, 2000; Bulletti *et al.*, 2001, Chapter 7), which are required to be investigated. However, some studies have produced evidence for an oocyte-derived Maturation Inhibiting Factor (MIF) and for a Maturation Promoting Factor (MPF) (Guraya, 1985, 1997a, 2000a; Sagata, 1996; Fauser *et al.*, 1999). A comparative study is required to be made of such factors in the oocytes of young and aged women. However, the acquisition of meiotic competence appears to depend on a genetical program developed by the oocyte itself but appears to be initiated by granulosa cells (GCs) at the start of follicle and oocyte growth. One important function of preovulatory follicular growth and maturation is, therefore, to form an oocyte which can undergo maturation and fertilization once ovulated from the follicle. Two patterns of acquiring meiotic competence can be found when various species of mammals investigated so far are compared (Wassarman and Albertini, 1994; Fauser *et al.*, 1999; Guraya, 2000a). The meiotic competence in rodents related to the time of antrum formation when oocytes almost have attained their maximum growth, but in pigs and cows, the meiotic competence is more closely related to the final growth of the oocyte and occurs in comparatively larger-sized follicles (Guraya, 1997a). These two patterns show comparative differences between short and long-cycling species. The acquisition of meiotic competence in the sheep takes place over a 24-hour period before ovulation and involves two phases (Moor and Gandolfi, 1987; Moor *et al.* 1992). The sheep oocyte during the first 6 hours phase does not show appreciable structural and synthetic alterations but instead apparently undergoes progamming by the GCs (Moor and Gandolfi, 1987; Moor *et al.*, 1992). During the second longer (18 hours) phase, most components of the oocyte undergo reorganization; GCs as pointed out do not play an important role during this period. The acquisition of meiotic competence, which forms at least a two-step process, appears to depend on FSH, possibly exerting its influence via stimulation of oestrogen secretion as discussed by Guraya (1997a, 2000a) and Suikkari *et al.* (2000). Since oocytes do not posses receptors for gonadotrophins and steriod hormones, somatic cells of the follicle play a role in development of meiotic competence (Wassarman and Albertini, 1994; Eppig *et al.* 1996; Guraya, 1997a, 2000a). Such studies carried out on the oocytes of other mammalian species are required to be carried out to determine the differences in process of oocytes in young and aged women for determining the effect of aging of primordial oocytes (Guraya, 1999; Chapter 1).

Recently, ovum maturation has been extensively studied in different mammalian species to have better insight into the regulatory mechanisms involved in it (Wassarman and Albertini, 1994; Guraya, 1997a, 2000a). But most of these studies have been carried out on the laboratory mammals and sheep. However, some studies have also been made of the ovum maturation in humans (Downs, 1996; Gosden *et al.* 1996; Szell *et al.*, 1996; Salha *et al.*, 1998; Fauser *et al.*, 1999; Wolf and Zelinski-Wooten, 2001) and mammals (Guraya, 1997a, 2000a), in which ovarian stimulation with exogenous gonadotrophins is likely to lead to the recovery of at least a few immature oocytes. If these are not brought to full maturity *in vitro* (Wolf and Zelinski-Wooten, 2001), they represent a loss of efficiency in the treatment cycle. Chen *et al.* (2000a) have suggested that a schedule to inject in *in vitro* matured oocytes in an intracytoplasmic sperm injection program cycles may generate more accessible embryos for fresh transfer or cryopreservations to increase the chance of pregnancy, although the embryo quality was relatively poor (Wolf and Zelinski-Wooten, 2001). The optional quality of both the oocyte and maturation medium are prerequisites for an undisturbed cytoplasmic maturation (Kruip *et al.*, 2000; Wolf and Zelinski-Wooten, 2001) showing that the *in vitro* maturation period does not compromise subsequent embryonic development (Smith *et al.*, 2000). Anderiesz *et al.* (2000) have studied the regulation of human and mouse oocyte maturation *in vitro* with 6-dimethylaminopurine. A brief account of the various concepts developed from extensive *in vivo* and *in vitro* studies on ovum maturation in laboratory mammals, some domestic ruminants and human will be given here and for more details, a reference is made to several reviews (Guraya, 1985, 1997a, 2000a; Salha *et al.*, 1998; Fauser *et al.*, 1999; Wolf and Zelinski-Wooten, 2001). The relevance of results of all these previous studies will be discussed in relation to the better understanding of cellular and molecular aspects of ovum maturation in young and aged women.

3.1 NUCLEAR AND CYTOPLASMIC CHANGES

Both nuclear and cytoplasmic changes form the specific feature of oocyte maturation, which are required for normal fertilization and development of the egg (Yang, 1997; Guraya, 1997a, 2000a; Moor *et al.*, 1998; Heikinheimo and Gibbons, 1998; van den Hurk *et al.* 1999; Yen *et al.*, 1999; Fauser *et al.*, 1999; Wolf and Zelinski-Wooten, 2001). Nuclear maturation can occur in an oocyte in which cytoplasmic maturation is not yet complete or otherwise defective and this may result in some of the conception failures of embryonic wastage in humans and domestic ruminants while carrying out *in vitro* maturation and fertilization and embryo transfer (Sharma and Chowdhury, 1998; Wolf and Zelinski-Wooten, 2001). Various studies have demonstrated that the association of the oocyte with the somatic GCs is essential for the maintenance of meiotic arrest (Guraya 1997a, 1998a,b, 2000a). Besides *in vivo* preovulatory maturation of the oocyte, maturation of oocytes *in vitro* as determined by resumption of meiosis, can be obtained by removal of the oocyte from inhibitory effects of GCs. It is now well known that in some species, in response to the effects of circulating FSH and LH after removal from Graafian follicles and *in vitro* culture at the dictyate-stage the oocyte of the mammalian ovary starts undergoing meiosis (reductional cell division), resulting in the final maturation of the oocyte within the mature preovulatory follicle(s) (Guraya, 1985, 1997a, 2000a; Thibault *et al.*, 1987; Wassarman and Albertini, 1994; Eppig *et al.*, 1996; van den Hurk *et al.*, 1999; Fauser *et al.*, 1999; Wolf and Zelinski-Wooten, 2001). Such studies are required to be carried out on the oocytes of young and aged women for determining the differences in the final maturation of their oocytes (Tevelde *et al.*, 2000; Bulletti *et al.*, 2001). Salha *et al.* (1998) have suggested that the physiological trigger for meiotic resumption in the human oocyte is the surge of LH, but it can also occur spontaneously if oocytes are released from antral follicles and cultured *in vitro* (Fauser *et al.*, 1999; Wolf and Zelinski-Wooten, 2001). Such a *in vitro* system could open the door to a number of techniques to carry out basic physiological studies on follicular development and oocyte maturation as well as to study the effects of toxicological substances on oocyte maturation. Such a system could provide a source of human oocytes for *in vitro* fertilization (IVF) (Wolf and Zelinski-Wooten, 2001). This system will facilitate oocyte cryopreservation of surplus oocytes, thus avoiding the need for repeated super ovulation. A combination of immature oocyte cryopreservation for later maturation and IVF will provide the opportunity to establish oocyte banks and help overcome some of the practical and ethical problems that are currently shadowing the field of reproductive medicine. The aspiration of immature oocytes from antral follicles followed by their maturation *in vitro* is a potential alternative to hormonal stimulation of patients in IVF treatment (Moor *et al.*, 1998; Wolf and Zelinski-Wooten, 2001). Although relatively successful in a variety of animals, the production of fully viable human embryos by *in vitro* maturation is still unsatisfactory despite the use of a wide variety of culture protocols. But Moor *et al.* (1998) have suggested that the key to maturation and embryo viability *in vitro* resides in the follicle cell compartment rather than the oocyte. Follicle cells in culture possibly fail to provide the maturing oocyte with the necessary ordered set of instructive signals and nutrients required for the acquisition of developmental competence as discussed in Chapter 2. Different steroids, matrix metalloproteinases and growth factors are known to be involved in conferring viability on the maturing oocyte as will be discussed here. A systematic analysis of somatic signals from the preovulatory follicle is still required (Chapter 2).

Although meiosis starts to occur in response to stimulation of the follicular syncytium by the preovulatory gonadotrophin surge (Buccione *et al.* 1990; Wassarman and Albertini, 1994; Guraya, 1997a, 2000a; Heikinheimo and Gibbons, 1998; Fauser *et al.*, 1999). Such a situation can be mimicked *in vitro* by the addition of gonadotrophins to cultures of intact antral follicles in several species including human, or in cumuls oophorus (CO) complexes grafted to segments of the follicle wall (Wolf and Zelinski-Wooten, 2001). Cumulus cells (CCs) are observed to improve the *in vitro* maturation rate compared with untreated controls (Ocana-Quero *et al.* 1997). Response to gonadotrophins is mediated by CCs (Buccione *et al.*1990; Mathioli, 1992), which shows receptors for gonadotrophins, whereas the oocyte does not. In response to the initial stimulation somatic cells of the follicle make a factor that induces known meiotic resumption of the oocyte in a paracrine fashion (Buccione *et al.*1990; Eppig, 1994; Eppig *et al.* 1996; Fauser *et al.*, 1999). This follicle cell factor result in a decrease in intracellular cAMP within the oocyte and commitment to the first division occurs, as will be discussed later. Such studies are required to be made on the oocytes of young and aged women to find out if any differences occur in regard to the nature and production of factor (s) by the somatic cells of the follicle, which influence resumption of meiosis in them (Tevelde *et al.*, 2000; Bulletti *et al.*, 2001).

The stimulus produced by the ovulatory surge of LH acting on a fully developed preovulatory follicle results in resumption of both meiosis or germinal vesicle breakdown (GVBD) by the oocyte (Mattioli, 1992; Albertini *et al.* 1993; Heikinheime and Gibbons, 1998; van den Hurk *et al.* 1999; Fauser *et al.*, 1999). Various studies have shown the requirement of LH for ovulation and resumption of meiosis by hCG and LH (Guraya, 1985, 1997a, 2000a; Heikinheime and Gibbons, 1998; Driecoll *et al.* 2000). Anderiesz *et al.* (2000a,b) have investigated the comparative effects of recombinant hCG on human, bovine and murine oocyte meiosis, fertilization and embryonic development *in vitro*. These data provide support for the responsiveness of human and bovine oocytes to gonadotrophins *in vitro* but not of mouse oocytes, suggesting the need to consider variations in the relative concentrations for optimization of oocyte developmental competence. Antibodies against LH inhibit resumption of meiosis and ovulation, inspite of the fact that a FSH surge occurs at the same time (Billig *et al.*1988). The results of various studies indicate that the LH effect on the oocyte mediates via GCs as the oocyte itself does not develop receptors for LH (Buccione *et al.* 1990; Guraya, 2000a). The final evidence of a normal meiosis is produced by the fertilization of the egg, implantation and a normal offspring (Guraya, 2000a; Wolf and Zelinski-Wooten, 2001; chapters 5 and 6). It will be rewarding to produce evidences for normal meiosis at the cellular and molecular levels in the oocytes of young and aged women by making comparative studies of meiosis, fertilization of the egg, implantation and a normal offspring (Fauser *et al.*, 1999; Wolf and Zelinski-Wooten, 2001). However, with initiation of meiosis, the chromosomes pass rapidly through diakinesis, metaphase I, a very short anaphase I and telophase I, and on the metaphase II of meiosis (Fauser *et al.*, 1999). At this stage, movements of chromosomes are again arrested and the ovulated egg awaits fertilization (Guraya, 1985, 1997a, 2000a; Moor *et al.*, 1992; Fauser *et al.*, 1999; Wolf and Zelinski-Wooten, 2001). A comparative study of behaviour of chromosomes during various stages of meiosis in the oocytes of young and aged women is required to be made to know if some differences occur (Tevelde *et al.*, 2000; Bulletti *et al.*, 2001).

Divergent views have been expressed about the molecular mechanisms of effects of gonadotrophins in the induction of meiosis (Fauser *et al.*, 1999). But recent studies have shown that there exists a cell to cell communication between the oocyte and its associated GCs, which is essential for regulation of meiotic mechanisms about which our knowledge is still incomplete, not only for mammals in general but also for human (Buccione *et al.* 1990; Albertini *et al.*, 1993; Eppig *et al.*, 1996; Guraya, 1997a, 2000a). The expansion of the CCs adjacent to the oocyte constitutes the mechanism whereby coupling is terminated. In a few mammalian species, the dictyate oocytes may undergo normal ovulation but mature in the oviduct, but such a situation is required to be demonstrated for young and aged women (Tevelde *et al.*, 2000; Bulletti *et al.*, 2001). The spherical germinal vesicle, which is either placed in the center of oocyte (mouse) or migrates to the priphery (in most of other mammals) shows either a single or more rarely double nucleoli (Thibault *et al.* 1987; Wassarman and Albertini, 1994; Guraya, 2000a). The chromatin masses in the GV of some mammals condense along the inside of the nuclear envelope projecting towards the nucleolus. A perinuclear chromatin crown is also developed in the oocytes of some mammals. It will be rewarding to make comparative study of chromatin masses in the oocytes of young and aged women during meiosis. The GV of a truly preovulatory oocyte prior to LH surge flattens against the plasma membrane. Whatever the position of the GV, the rupture of the nuclear envelope or germinal vesicle breakdown (GVBD) occurs in response to the Meiotic Promoting Factor (MRF), which still requires to be identified precisely at the molecular level in mammals and human (Albertini *et al.*, 1993; Fauser *et al.*, 1999). The details of GVBD are investigated in oocytes of other mammals (Guraya, 2000a) but not for young and aged women.

A sequence of well-defined nuclear and cytoplasmic maturation alterations occur during meiosis, which includes (1) disappearance of nucleolus, (2) dissolution of nuclear envelope, (3) formation of microtubules organizing centres (MTOCs), (4) inhibition of nuclear RNA synthesis, (5) mixing of nucleoplasmic and cytoplasmic components, (6) alignment of chromosomes on the metaphase I spindle, (7) separation of homologous chromosomes, (8) extrusion of the first polar body and (9) finally the arrest of meiosis at the metaphase II, i.e., the second meiotic arrest (Fig. 14) (Guraya, 1985, 1997a, 2000a; Heikinheime and Gibbons, 1998). The results obtained by Calarco (1995) in regard to polarization of mitochondria during the maturation of mouse oocyte should help us to understand organelle localization during mammalian and human oocyte maturation (Figs. 9A, 9B and 15) as well as during maturation of oocytes in young and aged women.

The dynamic alterations occur in membrane ultrastructure, intracellular organelles and nuclear structures and in cell cycle proteins and kinases during oocyte maturation, activation and fertilization (Yang, 1997; Fauser *et al.*, 1999). Microtubules polymerize in the intranuclear and interchromosomal spaces simultaneously with a clear cut GVBD. The precise regulation of these changes, which are well demonstrated with electron microscopy (Thibault *et al.*, 1987), are still required to be understood more precisely at the molecular level (Fauser *et al.*, 1999). Our information about the localization and nature of cytoskeletal elements during oocyte growth and maturation in mammals is progressively increasing as reviewed by Thibault *et al.* (1987). This pertains to the distribution of actin, tubulin and to the MTOCs, while other components belonging to keratin family are just emerging. Teruda et al. (1995) have demonstrated microfilaments during oocyte maturation of golden hamster, which are believed to play a key role in oocyte cytoplasmic maturation but the details need to be established. However, such studies on cytoskeletal elements are still to be carried out during oocyte maturation in the young and aged women (Fauser *et al.*, 1999). Protein phosphatases regulate mitogen-activated protein kinase activation and microtubules organization during rat oocyte maturation (Zernica-Goetz *et al.* 1997).

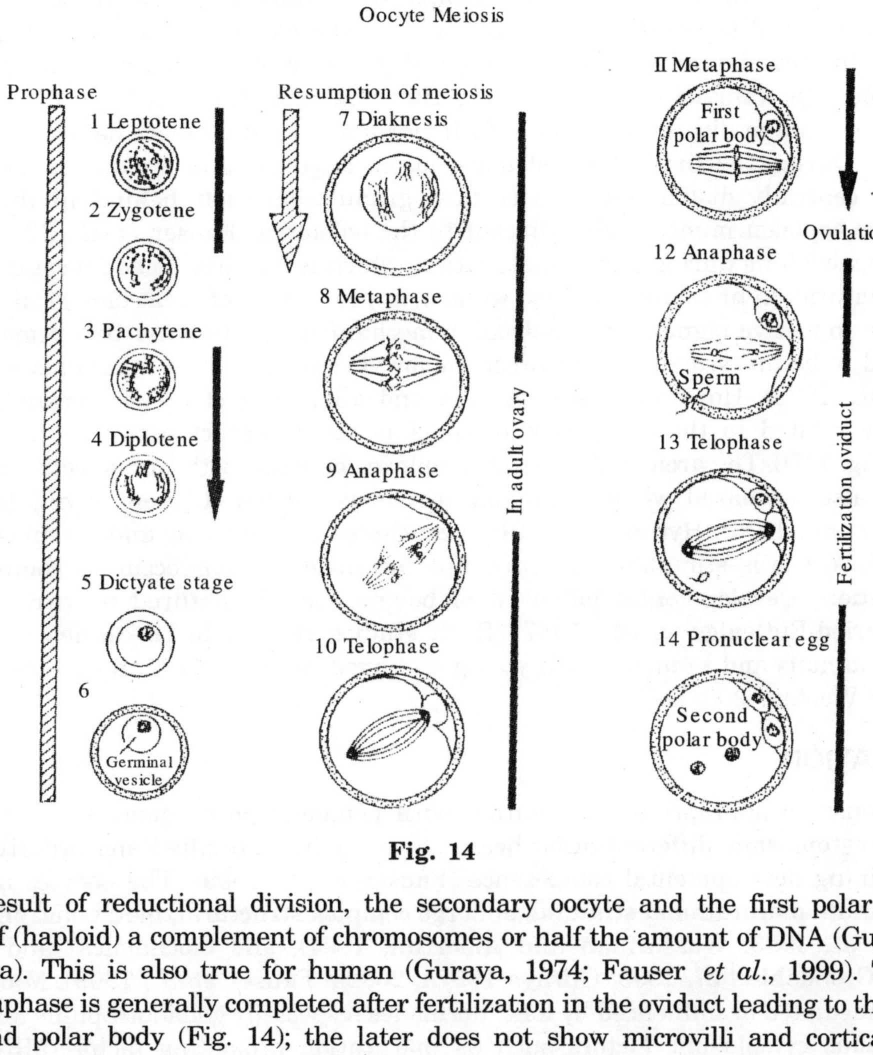

Fig. 14

As a result of reductional division, the secondary oocyte and the first polar body each contains half (haploid) a complement of chromosomes or half the amount of DNA (Guraya, 1985, 1997a, 2000a). This is also true for human (Guraya, 1974; Fauser *et al.*, 1999). The second meiotic metaphase is generally completed after fertilization in the oviduct leading to the extrusion of the second polar body (Fig. 14); the later does not show microvilli and cortical granules present in the first polar body as will also be discussed in Chapter 5. Even the behaviour of nuclear complement in both the polar bodies differs greatly and its regulation still remains to be determined. Various cytological and molecular changes occurring in the cytoplasmic organelles and other components are indicative of cytoplasmic maturation (Wassarman and Albertini, 1974; Heikinheimo and Gibbons, 1998; Fauser *et al.*, 1999; Guraya, 2000a).

Corresponding to nuclear changes, there occur rapid alterations in the spatial distribution of ooplasmic components such as the mitochondria, endoplasmic reticulum, microtubules, Golgi bodies, cortical granules (moving to the periphery), and lysosomes, and surface components of the oocyte during the period immediately preceding GVBD (Guraya, 1985, 1997a, 2000a; Hyttel *et al.*, 1986; Thibault *et al.*, 1987; Moor and Gandolfi, 1987; Moor *et al.*, 1992) indicating the possible presence of some biophysical forces during ovum maturation which are required to be determined not only for mammals but also for human. The redistribution of organelles appears to be mediated by microtubules radiating from MIOCs that develop during the GV to metaphase I transition. Due to some forces, the mitochondria get localized in the peripheral ooplasm whereas elements of rough endoplasmic reticulum, membrane-bound vesicles and cortical granules are placed more centrally in GV oocytes of domestic animals (Thibault *et al.*, 1987; Calarco, 1993). After GVBD, the rough endoplasmic reticulum disappears, and mitochondria and lipid droplets lie in clusters (Kruip *et al.*, 1983). During the final phase of maturation, aggregates of organelles are dispersed and become centrally distributed; the cortical granules are left behind in the relatively organelle-free ooplasm immediately adjacent to the oolemma (Fauser *et al.*, 1999). Such an intracellular modelling during oocyte maturation, which is required to be investigated during the oocyte maturation in young and aged women, appears to be of great biological importance although the inductive signals and molecular mechanisms underlying this remodelling are still required to be investigated more precisely for various groups of mammals and humans (Fauser *et al.*, 1999). However, the migration and alignment of cortical granules beneath oolemma are related to the disruption of the junctional contact between the oocyte and granulosa (Fig. 13D). The area of junctional complexes between both corona cells and cumulus and GCs is also decreased when ovum maturation is started (Larsen *et al.*, 1987; Eppig 1994; Eppig *et al.*, 1996). Hyttel *et al.* (1986) concluded that nuclear and cytoplasmic oocyte maturation in cattle is stimulated *in vitro* but certain deviations occur as compared to *in vivo* maturation. Developmental potential of bovine oocytes matured in vitro or *in vivo* differs (Leibfried-Rutleidge *et al.*, 1987). Such studies remain to be carried out for more species of mammals and humans, both young and aged women (First and Fields 1994; Wolf and Zelinski-Wooten, 2001).

3.2 REGULATION

In response to appropriate stimulation with gonadotrophins meiosis is initiated and nuclear and cytoplasmic differentiation begins to occur in a specified and orderly way with oocytes attaining developmental competence (Fauser *et al.*, 1999). The oocytes in response to an appropriate maturational stimulus undergo complex structural, metabolic, physiological (Albertini *et al.* 1993; Wassarman and Albertini, 1994), and biochemical and molecular alterations (Grondahl *et al.* 1995; Guraya 1997a, 2000a; Fauser *et al.*, 1999). Many of these changes are mediated or influenced by ions, hormones (especially gonadotrophins and steriods) and endogenous cytoplasmic factors such as maturation promoting factor (MRF), growth factors etc. (Guraya, 1997a, 1998b, 2000a; Sirotkin *et al.* 1998; Moor *et al.*, 1998; Fauser *et al.*, 1999); maximal sensitivity of Ca^{2+} release occurs in the final phases of oocyte maturation (Carroll *et al.* 1996; Fauser *et al.*, 1999). The findings of Goud *et al.* (1999b) coincide with and thus possibly represent the dynamic changes occurring in the cellular. Ca^{2+} release systems through oocyte maturation, fertilization and early embryogenesis. Thus, type I inositol 1, 4, 5 triphosphate receptors are likely to play a role during these stages of early development in the human Cobo *et al.* (1999) after studying maturation *in vitro* of human oocytes from unstimulated cycles have suggested the importance of retrieving immature oocytes before follicular selection, and define the conditions for the first stage in the use of immature oocytes (Wolf and Zelinski-Wooten, 2001).

Extensive studies made so far have clearly demonstrated the functional significance of intrafollicular somatic cell interaction or functional involvement of GCs in the regulation of progressive growth and differentiation of the oocyte (Eppig 1994; Eppig *et al.*, 1996; Guraya, 1997a, 1998b, 2000a) but also in mediating meiotic maturation and establishing required conditions for normal fertilization and embryogenesis (Buccione *et al.*, 1990; Grondahl *et al.*, 1995; Hyttel *et al.* 1997; Fauser *et al.*, 1999; Wolf and Zelinski-Wooten, 2001). In this regard, cellular and molecular aspects of communication or interactions between GCs and oocyte are being studied to determine the nature of factors involved in the growth and maturation processes of oocytes (Eppig, 1984; Eppig *et al.*, 1996; Guraya, 1997a, 1998b, 2000a; Fauser *et al.*, 1999; Chapter 2). It will be rewarding to carry out researches in these lines on the oocyte maturation in young and aged women (Tevelde *et al.*, 2000; Bulletti *et al.*, 2001) as our knowledge is also very poor about the transcription of genes specific to various factors involved in the growth and maturation of their oocytes as the human primordial oocytes undergo aging changes as discussed in Chapter 1. However, one current example of such a paracrine factor from GCs that affects oocyte growth in rodent is the product of steel factor (SLF). The close relationship of the KL producing GCs and the oocyte bearing the appropriate receptor for the same suggests an important functional relationship. Striking evidences are produced to show that this functional and symbiotic relationship between the oocyte and GCs is important for oocyte development and maturation within the follicle as it gets disturbed in the absence of/or following the loss of GCs during atresia (Guraya, 1985, 1997a, 2000a; Hinrichs and Williams, 1997; Chapter 2) or mechanical separation resulting into termination of the communication between GCs and oocyte (Buccione *et al.* 1990; Eppig, 1994; Guraya, 1997a, 1998b, 2000a; Wolf and Zelinski-Wooten, 2001). The control of various morphological, biochemical or molecular and physiological changes, which occur simultaneously in the interrelationships between the oocyte and the cells of the CO and corona radiata, or during the period preceding the completion of meiotic division, are still controversial (Guraya, 1985, 1997a, 2000a; Buccione *et al.*, 1990; Moor *et al.*, 1992; Wassarman and Albertini, 1994; Fauser *et al.*, 1999) and thus further studies on the cellular and molecular aspects need to be carried on not only in mammals but also in humans (in young and aged women). Intercellular communication, which forms a characteristic feature of the cumulus-oocyte complex during the growth of the oocyte as discussed in Chapter 2 is terminated near the time of ovulation in response to the LH surge (Guraya, 1997a, 2000a; Hyttel *et al.*, 1997) or follicle atresia (Hinrichs and Williams, 1997) and at the same time alterations occur in sulfated glycosaminoglycans. Actually the termination of the communication between the oocyte and corona cells is clearly related to the disruption of junctional contact between the oocyte and GCs thus reducing the entry of meiosis inhibitory factors into the oocyte and thereby inhibitory nuclear material. Meiotic arrest in mammalian oocytes during the first part of their growth phase is believed to be due to either the absence of essential cell cycle regulatory proteins, the presence of meiosis-arresting substance or both (Fauser *et al.*, 1999). At least three important meiosis inhibitors such as cAMP, peptide designated OMI (oocyte meiosis inhibitor), and the purine metabolite, hypoxanthine are believed to be involved in GC-mediated maintenance of meiotic arrest in mammals (Guraya, 1997a, 1998b, 2000a) as will be discussed here. The various experimental studies carried out by Dekel *et al.* (1988)

show that it is most likely that cytoskeletal elements participate in the mechanism of LH-induced oocyte maturation. The factors that cause disassembly of cytoskeletal elements, other than LH mimic LH action to uncouple the oocyte from the CCs and initiate oocyte maturation. GCs also produce chemical compounds (e.g., sulfated glycosaaminoglycans) which influence the CO surrounding the oocyte. With approachment of ovulation, the CO undergoes considerable expansion or mucification (Eppig *et al.*, 1996; Guraya, 1997a, 1998b, and 2000a) as it becomes embedded in the viscous secretion as will also be discussed in Chapter 4. Cumulus expansion occurs in response to the preovulatory LH surge. Bovine CC expansion does not depend on the presence of an oocyte secreted factor (Ralph *et al.*, 1995). But expansion of the CCs-oocyte complex needs a transient induction of hyaluronan synthesis by the CCs (Tirone *et al.*, 1997). The regulation of its synthesis is primarily regulated at the transcriptional level (Fulop *et al.*, 1997). The cumulus and mural GCs isolated from preantral to preovulatory stage follicles stimulated cumulus expansion enabling factor (CEEF) activity *in vitro* and autocrine secretion of CEEF by GCs is involved in the control of cumulus expansion *in vitro* and *in vivo* (Guraya, 2000a). Such studies are required to be carried out on the problems of cumulus expansion in young and aged women (Fauser *et al.*, 1999; Tevelde *et al.*, 2000; Bulletti *et al.*, 2001).

The serum shows a low molecular weight, heat stable factor (s) that acts on the oocyte to stimulate cumulus expansion (Dain *et al.*, 1995). The observations made by Downs *et al.* (1995) support the hypothesis that type II protein kinase A mediated cAMP-stimulated cumulus expansion and resumption of meiosis, while direct elevation of type I protein kinase A within the oocyte is responsible for maintaining meiotic arrest. Functions of protein kinase A and C pathways can modulate maturation of bovine oocytes *in vitro* (Rose-Hellekani and Bavister 1996). Prostaglandins are involved in paracrine control of cumulus expansion (Guraya, 2000a). The CCs cocultured with GCs or in a GCs conditioned medium undergo cumulus expansion. However, addition of cyclooxygenase inhibitors to the GC cultures stops the action of conditioned media on cumulus expansion, whereas they do not inhibit expansion of cumuli in response to hormonal stimuli. Hirashima *et al.* (1997) have suggested that the changed concentration of hyaluronan and inter-alpha trypsin inhibitor in the preovulatory ovaries may contribute to their important clinical characteristics including CO complex expansion. Chiou *et al.* (1995) have investigated the effects of the integrity of oocyte cumulus complexes and hormone combinations added to the media on bovine oocyte maturation and subsequent development *in vitro*. Mouse oocytes control hyaluronic acid formation and mucification by FSH-stimulated CCs (Salustri *et al.* 1990). The molecular basis of highly sensitive expansion response of CCs to FSH is still to be determined (Guraya, 1985, 1997a, 2000a). FSH not only stimulates the expansion of isolated CO, but promotes the uncoupling of the CCs from the oocytes as well, which is accompanied by degenerative alterations in their fine processes and junctions leading to decrease in the transmission of factors into maturing oocytes (Dekel *et al.*, 1988). Morphological differentiation of CO complexes is controlled by gonadotrophins (Kanayama *et al.*, 1990). Growth factors also play important roles in this regard (Guraya, 2000a). Some workers believe that the disruption of junctional complexes between the oocyte and corona cells also inhibits the entry of meiosis inhibitor into the oocyte and thereby starts nuclear maturation (Buccione *et al.*, 1990; Eppig *et al.*, 1996; Guraya, 1997a, 1998b) as will be discussed later on. This may be due to the effects of FSH induced on the CCs. The germinal vesicle stage oocyte function in the maintenance of CCs oocyte metabolic cooperativity as complete uncoupling is seen in those oocytes, which show complete GVBD.

The precise physiological functions of sulfated glucosaminoglycans (present in FF) in oocyte-cumulus changes (especially in cumulus expansion) and in ovulation are still to be determined more precisely. The sulfated glucosaminoglycans may inhibit precocious expansion of the CO in response to FSH indigenous to the antral follicle before the preovulatory LH surge. *In vitro* investigations have shown that CCs isolated from mice can induce expansion of CO within 24 hours, and this expansion promoting property appears to be due to diffusible factors released into the culture medium. This property of causing cumulus expansion is believed to be mediated by PGE_2, which may help in facilitating the expansion of cumulus around the eggs of domestic ruminants and human for *in vitro* fertilization and ova transfer programs (Guraya 2000a; Wolf and Zelinski-Wooten, 2001). It will be rewarding to make comparative studies on the role of PGE_2 in the expansion of cumulus around the eggs of young and aged women (Tevelde *et al.*, 2000; Bulletti *et al.*, 2001). The investigations of Fukui *et al.* (1986) have shown that evaluations of expansion and stretchability of CC mass in post-cultured oocyte form the useful criteria to assess their maturity as also supported by the comparative analysis of the polypeptide pattern of CCs during maturation of porcine cumulus oocyte complexes *in vivo* and *in vitro* (Guraya, 2000a). According to King *et al.* (1986), the nuclear maturation of porcine oocyte is more rapid *in vitro* than *in vivo.* Such studies are required to be made on the nuclear maturation in the oocytes of young and aged women for determining the effects of aging as the oocytes devoid of CCs or showing signs of vacuolation or degeneration have virtually no ability for nuclear maturation. The bovine oocytes matured and fertilized *in vitro* failed to develop to the 2-cell stage whereas oocytes matured *in vivo* underwent development to the 2-and 4-cell stage (Leibfried-Rutledge *et al.*, 1987, 1989). Oocytes that are matured develop to morula *in vivo.*

The functional association (for metabolic cooperativity) between the oocyte and cumulus is well known to play a significant role in maintaining meiotic arrest (Buccione *et al.*, 1990; Eppig *et al.*, 1996). Ohta *et al.* (1999) have suggested that the standard form of CD44 (a polymorphic and polyfunctional transmembrane glycoprotein) is expressed in human CC and membrana granulosa with polarity and may play an important role in oocyte maturation (Wolf and Zelinski-Wooten, 2001). Previous studies indicated that the breakdown of functional association between the oocyte and cumulus cells in response to gonadotrophin surge constitutes the physiological trigger of the resumption of meiosis (Guraya, 1985, 1997a, 1998b, 2000a).

It is now believed that the gap junctional communication within the ovarian follicle is involved in the resumption process of meiosis, but divergent views exist in this regard (Guraya 1997a, 1998b, 2000a; Moor *et al.*, 1998, Fauser *et al.*, 1999; Wolf and Zelinski-Wooten, 2001). Some workers believe that the decrease in gap junctional communication between the oocyte-CC complex and the membrana granulosa reduces the transport of meiosis-inhibiting substances to the oocyte and thus to the resumption of meiosis, others believe that gonadotrophins induce the production of a positive maturation inducting signals within the GCs cells; the transfer of this signal to the oocyte would need patent gap junctional communication (Guraya, 2000a). Further studies are required to solve this controversy. Comparative investigations will be very rewarding to define more precisely the nature and molecular mechanism of transfer of signals to the oocyte in different groups of mammals and humans. A deeper understanding of such signals or factors from a comparative point of view will be very helpful in the better understanding of their roles in the regulation of meiosis not only in mammals (de Lacet *et al.*, 1989) but also in young and aged women (Fauser *et al.*, 1999; Tevelde *et al.*, 2000; Bulletti *et al.*, 2001; Wolf and Zelinski-Wooten, 2001). However, the results of some experimental studies suggest the GVBD does not occur in response to the termination of coupling between GCs and the oocyte of the ungulates (Moor and Gandolfi, 1987; Moor *et al.*, 1992). But a possibility can not be ruled out that initiation of GVBD is controlled by a qualitative or quantitative change in a factor transported to oocyte via the intact coupling pathway (Buccione *et al.*, 1990; Eppig *et al.*, 1996; Guraya, 1997a, 1998b, and 2000a). The results of various investigations as summarized by Thibault *et al.* (1987) and Leibfried-Rutledge *et al.* (1989) show that the presence of GCs is absolutely essential at the start of the final oocyte maturation in sheep, bovine, pig and rabbit. The results of these studies need to be extended and confirmed for the oocytes of young and aged women (Wolf and Zelinski-Wooten, 2001). However, GCs regulate protein and/or polypeptide synthesis, which renders the ooplasm competent to assume normal cooperation with the male genome. But this does not hold good for rodents, suggesting species differences at the molecular level.

Spontaneous initiation of meiosis occurs in mammalian oocytes freed from GCs (reviewed by Guraya 1985, 1997a, 2000a). Such studies are required to be carried out for the young and aged women to determine the differences at the molecular level (Fauser *et al.*, 1999; Wolf and Zelinski-Wooten, 2001). However, Gougeon and Testart (1986) have made study of cyclic evolution in GVBD percentages in oocytes of human atretic follicles during the menstrual cycles, which indicate that, the removal of inhibition (due to atresia) exerted by the follicle itself on the GV is insufficient to induce the resumption of meiosis in the human oocyte. Specific induction is required, at least to obtain nuclear maturation. Normal meiosis can occur after the transfer of a germinal vesicle (GV) into a nucleated host oocyte (Zhang *et al.*, 1999; Wolf and Zelinski-Wooten, 2001). GV transfer may be a valuable research procedure that generates cell models to characterize the cytoplasmic nuclear interplay for cell cycle regulation, maturation and fertilization in the human oocyte; it also may be a potentially attractive alterations to oocyte donation.

During the recent years, the effects of gonadotrophins, steroid hormones, cyclic nucleotides, prostaglandins, Gn-RH, FF inhibitors, ions, electrolytes and various other factors including growth factors on ovum maturation in numerous species of mammals have been extensively investigated under *in vivo* and *in vitro* conditions (Guraya, 1985, 1997a, 2000a; Thibault *et al.*, 1987; Billig *et al.*, 1988; Leibfried-Rutledge *et al.*, 1989; de Lacet *et al.*, 1989; Magoffin and Erickson, 1994; Wassarman and Albertini, 1994; Tsafriri and Adashi, 1994; Yang-Bing Chen *et al.*, 1995; Youshimura *et al.*, 1996 a, b, c; Moor *et al.*, 1998; Fauser *et al.*, 1999; Driscoll *et al.*, 2000; Wolf and Zelinski-Wooten, 2001). Inhibin alpha-subunit biosynthesis is associated with normal oocyte and follicle maturation, but excessive alpha-inhibin is associated with poor embryo quality in human (Fujiwara *et al.*, 2000; Wolf and Zelinski-Wooten, 2001). None of the hormones analyzed were associated with oocyte or embryo quality. Chian *et al.* (1999) have suggested that the admistration of hCG 36 hours before harvesting of immature oocytes may improve the maturational and developmental competence of the oocytes and the pregrancy rates of unstimulated patients with polycystic ovary syndrome. In a further study Chain *et al.* (2000) have made prospective randomized study of hCG priming before immature oocyte retrieval from unstimulated women with polycystic ovarian syndrome. Following 24 hours of culture, 78.2 + 1-7.1% of oocytes in the non-hCG-primed group were matured in the hCG-primed group compared with 4.9 + 1-7.1% oocytes ($P < 0.01$). There were fertilization and cleavages in these groups no significant differences in the rates of oocyte fertilization and cleavages in these groups occur. Stevenson (1999) have investigated the influence of GnRH/Gn RH agonist on steroidogenesis and IVF outcome. There is observed lower incidence of stimulations because of premature LH surges, which happen sometimes during ovarian stimulation. The implications of a direct influence on E2 synthesis and its effects on oocyte maturation are required to be investigated. However, such studies are required to be carried out on the ovum maturation and embryo quality in young and aged women to demonstrate differences at the molecular level (Wolf and Zelinski-Wooten, 2001). However, culture conditions can affect meiotic regulation in cumulus-enclosed mouse oocytes (Downs and Mastropolo, 1997; Wolf and Zelinski-Wooten, 2001). Smoking is found to alter the meiotic spindle of oocytes, leading to chromosomal errors, which affect reproductive outcomes (Zenzes, 2000).

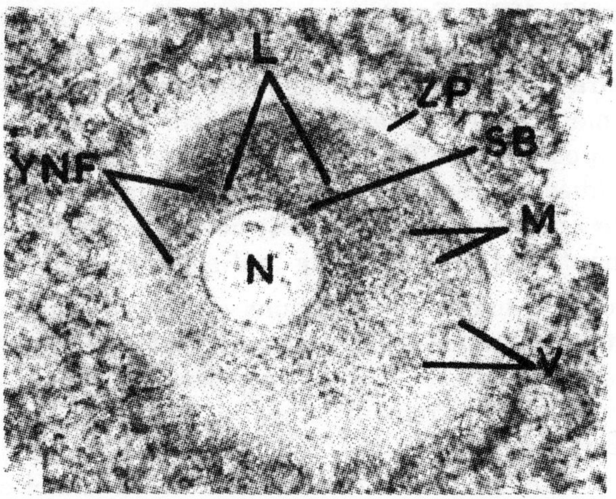

Fig. 15

Kito and Bavister (1997) have observed that gonadotrophins, serum and aminoacids can change nuclear maturation, cumulus expansion and oocyte morphology in hamster cumulus-oocyte complexes *in vitro*. IGFBP-3 inhibits oocyte maturation (Yoshimura *et al.*, 1996 a, b, c). Can and Albertini (1997) have described stage-specific effects of carbendazim (MBC) on meiotic cell cycle progression in mouse oocytes. The endothelial nitric oxide synthases-derived nitric oxide is observed to be the modulator of oocyte meiotic maturation (Jablonka *et al.*, 1996). The major objectives of all these studies were to demonstrate the nature of structural and chemical changes of regulatory mechanisms involved in ovum maturation at the molecular level. Different kinds of chemical factors are observed to function as oocyte maturation-inhibiting factor (OMIF). These include peptides, nucleotides, nucleosides, purines and steroids (Fauser *et al.*, 1999). Downs (19997b) has observed the involvement of purine nucleotide synthetic pathways in gonadotrophin induced meiotic maturation in mouse cumulus cell enclosed oocytes. Anderiez *et al.* (2000) after studying the regulation of human and mouse oocyte maturation *in vitro* with dimethylaminopurine (DMAP) have suggested that lengthening the prematuration growth phase, by temporarily inhibiting kinase activity with DAMP, does not directly improve oocyte developmental competence but provides a useful tool for further investigating meiotic and developmental related events *in vitro* by manipulating meiotic resumption.

Beevers *et al.* (1997) have discussed the endocrine and paracrine factors that affect the *in vitro* maturation of bovine oocytes as also discussed by Pawshe *et al.* (1998) for buffalo. Among all potential OMIFs, one or the other appears to perform the main function, depending not only on the species but also on the experimental designs. Thus, it becomes difficult to establish a general hierarchy among these factors and furthermore they also often function synergistically. Thus, results obtained during various investigations are still controversial and thus further studies are required to be carried out at the molecular level not only for mammals but also for young and aged women (Tevelde *et al.*, 2000; Bulletti *et al.*, 2001; Wolf and Zelinski-Wooten, 2001). The capacity of some drugs and hormones to stimulate meiosis must merely be considered pharmacological. Species variations are also recorded (Guraya, 1985, 1997a, 2000a; Eppig, 1987, 1994; Thibault *et al.*, 1998; Wassarman and Albertini, 1994; Tsafriri and Adashi, 1994; Eppig *et al.* 1996). Investigations have been made of maturation of slaughterhouse ovarian follicular oocytes of buffalo in culture and subsequent *in vitro* fertilization (Guraya 1997a, 2000a). Very little information is available about oocyte physiology and *in vitro* maturation and fertilization of the oocyte in the buffalo. Furthermore, *in vitro* matured buffalo oocytes, unlike cattle oocytes do not operate an effective block to polyspermy. Further studies are required to establish the culture length, which will give the highest maturation rate and meanwhile the best developmental competence of oocytes not only in domestic ruminants but also in young and aged women.

Actually, the results of many investigations on *in vitro* maturation are difficult to extend and confirm as the conditions of the biological material used are not sufficiently well defined either at the start or the end of the experiment (Salha *et al.*, 1998; Wolf and Zelinski-Wooten, 2001). However, Thibault *et al.*, (1987) suggested that theoretically the oocytes from preovulatory follicles are the only ones suitable for the study of molecular aspects of inhibitors. The cytological quality of the granulosa and/or corona cells is not much studied although meiosis may resume in atretic follicles (Billig *et al.*, 1988; Buccione *et al.*, 1990; Hinricks and Williams, 1997), leading to a misinterpretation of the inhibitory effect of the substance studied. A superficial examination of the GV or of its disappearance (GVBD) is not sufficient to determine the cytological effect of the meiosis inhibitors used. However, the GVBD appears to be a proper criterion when the investigation is concerned with the identification of molecular mechanisms controlling the resumption of meiosis (Moor *et al.*, 1992; Fauser *et al.*, 1999). But it should not be considered indicative of fertilizability of the ovum or of its competance to form a viable embryo (Tsafriri and Adashi, 1994; Guraya, 2000a; Wolf and Zelinski-Wooten, 2001).

Various studies have demonstrated the presence of some factors involved in the maintenance of meiotic arrest, but their precise nature is still controversial even the cause of their inactivation after the preovulatory gonadotrophin surge is required to be determined (Thibault *et al.*, 1987; Billig *et al.*, 1988; Leibfried-Rutledge *et al.*, 1989; Fauser *et al.*, 1999). At present, at least three important inhibitors such as the cAMP, peptide designated OMI (oocyte meiosis inhibitor) and the purine metabolite, hypoxanthine are suggested to be involved in the maintenance of meiotic arrest (Guraya, 1985, 1997a; Thibault *et al.*, 1987; Billig *et al.*, 1988; Leibfried-Rutledge *et al.*, 1989; Buccione *et al.*, 1990; Albertini *et al.*, 1993;

Tsafriri and Adashi, 1994; Downs, 1996, 1997; Fauser *et al.*, 1999). A comparative study is required to be made of these inhibitors in young and aged women to determine any difference in the mechanism of maintenance of meiotic arrest in them (Tevelde *et al.*, 2000; Bulletti *et al.*, 2001). However, the interruption of GCs oocyte communication in response to hormone (especially gonadotrophins) is believed to result in a decrease of cAMP as a physiological regulator of meiosis in mammalian oocytes (Guraya, 1985, 1997a, 1998a, b, 2000a). The presence of OMI was developed from the reports that GCs, follicular GCs extracts and media in which GCs are previously cultured and a filtrate of FF can inhibit meiotic maturation of isolated cumulus-oocyte complex in a culture (Wolf and Zelinski-Wooten, 2001). OMI is not species specific and is produced by somatic cells of the follicle, released into the FF and causes maintenance of meiotic arrest in the oocyte (Guraya, 1985, 1997a, 2000a; Fauser *et al.*, 1999). This OMI is a small molecular weight peptide (Tsafriri *et al.*, 1987). There is very strong evidence for the participation of purines, especially guanyl compounds in the maintenance of meiotic arrest *in vivo* (Leihfried-Rutledge *et al.*, 1989; Downs, 1996; Eppig *et al.*, 1996; Fauser *et al.*, 1999). But it is still not known whether production of these compounds *in vivo* involves hypoxanthine. By using cumulus-oophorus complexes from bovine ovaries, a transient inhibition of meiotic resumption by hypoxanthine and adenosine or their combination is demonstrated. Bovine oocytes from early antral follicles are observed to grow to meiotic competance *in vitro* in response to the effect of FSH and hypoxanthine (Harada *et al.*, 1997). Downs (1997) has discussed the various aspects of hypoxanthine regulation of oocyte maturation in mouse especially in relation to use of hypoxanthine phosphoribosyl transferase deficient animals (Downs, 1996). The role of hypoxanthine in the regulation of oocyte maturation is required to be determined with the aging of women as it is found in the FF (Guraya, 2000a).

The specific effects of both OMI and the purines are reversible and fulfil some of the criteria for a physiological regulator of oocyte maturation (Eppig, 1994; Billig *et al.*, 1988; Tsafriri and Adashi, 1994; Guraya, 2000a), hypoxanthine occurs in variable amount in the human and bovine FF. The observed synergism of cAMP and OMI in inhibiting spontaneous maturation *in vitro* of isolated oocytes shows that interaction of cAMP and follicular OMI is of great importance in the control of oocyte maturation (Wassarman and Albertini, 1996; Tsafriri and Adashi, 1994; Eppig, 1996; Guraya 2000a), further supporting the concept of multifactorial regulation of initiation of meiosis which still remains to be determined more precisely at the molecular level not only in human but also in various other mammalian groups (Fauser *et al.*, 1999). However, Sirotkin *et al.* (1988) have investigated the effect of follicular cells, IGF-I and tyrosin kinase blockers on pig oocytes *in vitro*. IGF-I stimulated meiotic maturation of both cumulus enclosed and cumulus-free oocytes. Neither of the tyrosin kinase blockers changed the stimulating effect of IGF-I. Intra-ovarian regulation of follicular maturation is modulated by growth factors (Fauser *et al.*, 1999), which are important local factors as discussed in detail by Guraya (2000a) IGF-I stimulates the meiotic maturation of follicle-enclosed oocytes in vitro via the IGF-I receptors (Voshimura, 1998; Wolf and Zelinski-Wooten, 2001). IGFBP significantly inhibits gonadotrophins induced ovulation and oocyte maturation by neutralizing endogenous by produced IGF-I, suggesting the important role of IGF-IGFBP system in the processes of follicular development, oocyte maturation and ovulation. Epidermal growth factor (EGF) combined with recombinant hCG improves meiotic progression in mouse follicle enclosed oocyte culture (Smitz *et al.*, 1998). Goud *et al.*, (1998) have suggested that the supplementation of the maturation medium with EGF and maintenance of the cumulus during culture improve the nuclear and cytoplasmic maturation of human oocytes *in vitro*. Values of transfer in the FF are found to be highly correlated with those in the serum, suggesting that the small contributions made by its localized synthesis in the granulosa cell may be important or some yet unknown mechanism in follicle for oocyte maturation (Briggs *et al.*, 1999). Reeka *et al.*, (1998) have studied the immunohistochemical localization TGF-alpha, EGF and EGF receptor in 18 human ovaries. These observations support the participation of TGF/TGF-alpha in follicular maturation. Furthermore, the presence of TGF-alpha in FF and the simultaneous absence of EGF suggests that TGF-alpha plays a more pronounced role than EGF in oocyte maturation during late follicular phase.

It is possible that the ovum maturation may be induced by a coordinated series of events, which include both a decrease in the generation or transfer of GVBD arresting factors and the production of some positive signals by GCs that promote maturational processes. Actually, positive signals are believed to provide a more precise control of maturational processes than a simple withdrawl of maturation-arresting factors. The pituitary gonadotrophins could generate a positive signal within mammalian CCs which will be capable of stimulating GVBD in the continuous presence of inhibitory factors (Downs et. *al.*, 1988). The positive signals may be some steroid hormones (progestins) produced by the GCs and CCs as demonstrated for the follicle cells of fish and amphibians (reviewed by Jalabert *et al.*, 1991) as well as growth factors (Hirshfield, 1989; de Laat *et al.*, 1989). The precise identification, chemical characterization and mechanism of action of putative mammalian maturation promoting signals forms most critical area for a better understanding of the mechanisms that start maturation and of the processes that promote the acquisition of potential for fertilization and embryogenesis (Leibfried-Rutledgo *et al.* 1989; Buccione *et al.*, 1990; Fauser *et al.*, 1999). Although contradictory results are reported in the literature the general consensus is that the cumulus oophorus forms the 'threshold' by which the controls of meiosis must traverse to exert their effect in the meiotically competent oocyte (Buccione *et al.*, 1990; Eppig *et al.*, 1996).

Some studies also reveal the possibility of the involvement of SRIF (Mori *et al.*, 1985) and antimullerian hormone (AMH) (Takahashi *et al.*, 1986a) in maintaining meiotic arrest (Billig *et al.*, 1988; Jalabert *et al.*, 1991). AMH is localized in the cytoplasm of GCs and FF of sheep and rat Graffian follicles (Takahashi *et al.*, 1986b; reviewed by Jalabert *et al.*, 1991). But the maturation-inhibiting function of AMH is controversial, possibly due to differences in the methods of preparation and assay, which may produce artifacts (Guraya, 2000a). Inhibin produced by the GCs and present in the FF (Guraya 2000a) is also observed to initiate spontaneous GVBD *in vitro* in both cumulus-enclosed and denuded rat oocytes. Inhibin A and activin A play important roles during the final stages of oogenesis including oocyte maturation (Stock *et al.*, 1997; Fauser *et al.*, 1999). But the molecular mechanism of their action remains to be determined more precisely. The precise roles of inhibin A and activin A are required to be investigated in women of all age groups (Fauser *et al.*, 1999).

OMI produced by the GCs and stored in the FF of developing follicles is believed to inhibit the resumption of meiosis. But divergent views exists about the presence of an inhibitor in the FF which inhibits the maturation of oocyte (Guraya 2000a). Fleming *et al.*

(1983) using porcine FF and rat oocytes did not observe any inhibitory action upon oocyte maturation which is independent of the nature of follicular stimulation, the diameter of the follicle from which the oocyte is obtained and irrespective of the source of fluid or its filterate from diverse follicles and different females. Oocyte maturation is inhibited when a protease inhibitor is added to fluids, this inhibitor is sometimes used to preserve the potential of the maturation inhibitor in FF. From these observations, the authors have questioned the concept of a specific inhibitor of oocyte maturation within the FF. Factor(s) in porcine FF is observed to inhibit the induction of cumulus expansion of oocyte-cumulus complexes cultured *in vitro* (Kim *et al.*, 1996). The results of some other investigations involving co-culture systems negate the function of GCs or FF inhibitor in spontaneous inhibition of nuclear maturation (reviewed by Leibfried-Rutledge *et al.*, 1989; Kim *et al.*, 1996). The efficiency of FF as a supplement to the maturation medium to improve cytoplasmic maturation seems to vary with follicle quality but not size (Carolan *et al.*, 1996). However, the addition of 10% FF or foetal calf serum (FCS) to the maturation media generally results in a similar blastocyst yield with no additional improvement when media is supplemented with FCS and FF. Various papers and reviews provide the details of recent advances in factors affecting oocyte maturation and subsequent development (First and Fields, 1994; Zhang *et al.*, 1995; Kim *et al.*, 1995; Funahashi *et al.*, 1997; Kanitz *et al.*, 1997; Mogas *et al.*, 1997; Fauser *et al.*, 1999; Yen *et al.*, 1999; Wolf and Zelinski-Wooten, 2001). But studies are still required to be made to determine the quality of oocytes and factors affecting oocyte maturation and subsequent embryogenesis in women of variable age groups as (Bulletti *et al.*, 2001; Tevelde *et al.*, 2001) the primordial oocytes are observed to undergo aging changes (Guraya, 1999, 2000a Chapter 1).

The results of various studies have demonstrated that the meiotic arrest of the oocyte in the follicle is now maintained by a steady supply of cAMP which is transported to the oocyte from the cumulus by the process of intercellular communication (Guraya, 1985, 1997a, 2000a; Thibault *et al.*, 1987; Billing *et al.*, 1988; Leibfried-Rutledge *et al.*, 1989; Buccione *et al.*, 1990; Mattioli, 1992; Wassarman and Albertini, 1994; Tsafriri and Adashi 1994; Downs *et al.*, 1995; Heikinheimo and Gibbons, 1998; Fauser *et al.*, 1999). The interruption of cumulus oocyte communication in response to gonadotrophins may result in a decrease in cAMP concentrations in the oocyte and the onset of meiotic maturation, indicating the role of cAMP as a physiological regulator of meiosis in mammalian oocytes (Billig *et al.*, 1998; Guraya, 2000a). The decrease in cAMP has also been attributed to an increased activity of phosphodiesterase enzyme or a decrease in adenylate cyclase activity in the oocyte. If high concentrations of cAMP maintain meiotic arrest in the oocyte, it shows that intrafollicular or intraoocytic inhibitors of meiosis must eventually be involved in controlling the concentrations of oocytic cAMP. The plasminogen activator formation by porcine oocyte CC complexes *in vitro* is influenced by protein kinase A and C and kinase inhibitors during oocyte maturation (Kim and Menino, 1995). Inhibition of intracellular phosphatases also induces novel plasminogen activator.

The decrease in cAMP concentrations may also be due to an enhanced activity of the phosphodiestrase enzyme (PDE) or a decrease of adenylate cyclase (AC) activity in the oocyte as already studied (Guraya, 2000a), however, the presence of AC activity in the oocyte is not clear (Billig *et al.*, 1988; Fauser *et al.*, 1999). Unfortunately, at present no information is available on the relationship between cAMP levels and oocyte maturation as well as on the influences of phosphodiestrase inhibitors on oocyte in domestic ruminants and humans (Guraya, 1997a, 2000a; Fauser *et al.*, 1999; Wolf and Zelinski-Wooten 2001). It is still to be investigated whether the increase in oocyte cAMP is actually the result of transport of cAMP from CCs or of the transfer of the factors that promote the formation of cAMP within the oocyte as cAMP precursersors stimulators of oocyte adenylate cyclase (Wassarman and Albertini, 1994). The oocytes of several species can produce their own cAMP, whether they can produce it in sufficient quantities to maintain oocytes in meiotic arrest under physiological conditions is still required to be determined. However, the results of various studies show that the influx of cAMP is arrested and PDE in the oocyte possibly lowers the cAMP concentrations. A decrease in oocyte cAMP concentrations is of critical significance for oocyte maturation and that this decrease causes decreased phosphorylation of certain proteins by the cAMP dependent protein kinase (PK-A). Both stimulators of the cAMP-PKA-system and those stimulators phosphodiester turnover and PKC can under experimental conditions induce resumption of meiosis (Leibfried-Rutledge *et al.*, 1989; Fauser *et al.*, 1999). The simlest model claims basal concentrations of cAMP formation within the rat GCs, producing sufficient cAMP to maintain elevated intraoocyte levels after transfer via gap junctions (Dekel, 1998). Another model proposes adenosine augmentation FSH-dependent cAMP productions in GCs for maintenance of intraoocyte level of this adenyl compound (Behrman *et al.*, 1988). The murine model lays emphasis on a synergism between adenosine and hypoxanthine that ultimately modulates oocyte maturation (Downs, 1996; Guraya, 2000a). The promotion of generation of guanosine compounds by hypoxanthine rather than hypoxanthine alone is believed to enhance levels of cAMP within the oocyte; the site of action of the purines is on the GCs of the follicle. Evicence has been produced for cumulus-facilitated uptake of hypoxanthine, adenosine and guanosine by mouse oocyte (Leibfried-Rutledge *et al.*, 1989).

The concept of decrease in oocyte cAMP levels in relation to resumption of meiosis as demonstrated for the mouse and some other mammals does not apply to meiotic regulation of sheep oocytes as nuclear changes precede junctional disruption in this species, and cAMP concentrations do not decrease at the initiation of maturation but are enhanced significantly at this time (Moor and Gandolfi, 1987; Moor *et al.*, 1992). This negates the role of cAMP in the inhibition of meiosis in sheep oocytes and so for no other inhibitory factor is observed in this species. Whether cAMP or OMI or both, function as the inhibitory signal at the level of the oocyte in other mammalian species as well as in human still remains to be determined more precisely at the molecular level (Billig *et al.*, 1986; Wassarman and Albertini, 1994; Fauser *et al.*, 1999). Even the molecular mechanisms and routes of intercellular transfer of such regulatory molecules between GCs and oocytes are still required to be determined more precisely. How gap junctions formed between the oocytes and GCs plasma membranes (Chapter 2) are believed to provide direct access to the oocyte cytoplasm of molecules that can not be transported directly across the cell membrane (Buccione *et al.*, 1990; Eppig *et al.*, 1996; Guraya 2000a). Such intercytoplasmic channels appear to function as conduits for transport of OMI and cAMP from either the CCs or surrounding follicular environment to the oocyte cytoplasm, where they may be involved in the inhibition of meiosis reinitiation. The inhibitory effect of OMI upon meiosis appears to be exerted; at least partially, through the mediation of CCs (Leibrfied-Rutledge *et al.*, 1989; Buccione *et al.*, 1990). Moor and Gandolfi (1987) have suggested that signals from the GCs, functioning as distance activator stimulate the adenylate cyclase system, leading to the subsequent increase of cAMP concentrations in the oocyte. Intracellular concentrations of calcium and cAMP are also believed to function synergistically, through a calmodulin-dependent step, to regulate meiotic maturation of mammalian oocytes (Leibfried-Rutledge *et al.*, 1989; Wassarman and Albertini, 1994; Fauser *et al.*, 1999). Ca^{2+} is found to participate in the progression of meiosis and that 1, 4, 5, inositol triphosphate receptor is considered to be responsible for the majority of Ca^{2+} release during oocyte maturation and fertilization (He-Changli *et al.*, 1997). The results of some further investigations support the hypothesis that a factor of CCs origin causes inhibition of oocyte maturation by a cAMP-dependent process (Billig *et al.*, 1988; Buccione *et al.*, 1990; Eppig *et al.*, 1996; Guraya 2000a). The relationship if any between OMI-and the putative inhibiting factor possibly purine nucleotides is still required to be determined (Billig *et al.*, 1988). The purification and chemical characterization of OMI are required to reveal the precise nature of biochemical and cellular mechanisms controlling the preovulatory resumption of meiosis. Bagger *et al.* (1988) have suggested that in the mouse, CCs synthesize a meiosis inducing or activating substance which stimulates resumption of meiosis and that this effect is possibly mediated via the phosphatidylinositol pathway (Guraya 2000a). Gn-RH can induce resumption of ovum maturation, CC dispersion and mucification (Jalabert *et al.*, 1991; Tsafriri and Adashi 1994). Its action on ovum maturation involves protein kinases. Thus, phospholipase C, phorbol ester and diacylglyceral can motivate resumption of meiosis in follicle-enclosed oocytes and synergise with Gn-RH in this action (Guraya, 2000a). Inhibition of the lipoxygenase pathway of arachidonic acid inhibits the resumption of meiosis induced by Gn-RH, but not by LH, indicating the involvement of this pathway in the mediatory action of Gn-RH on ovum maturation (Tsafriri and Adashi, 1994).

Heikinheimo and Gibbons (1998) have reviewed the previous literature to outline the current understanding on the molecular mechanisms governing various stages of oocyte maturation transition from maternal to embryonic control and the initial steps of pre-embryo development (Fauser *et al.*, 1999). There occurs a decrease in the intracellular concentrations of cAMP. This and several subsequent steps of meioses are controlled by the M-phase promoting factor (MPF). An oocyte specific protein kinase, c-mos, plays an important role in upregulating the activity of MPF at various stages of final oocyte maturation. Several lines of evidence suggest that the proper function of the c-mos-MPF system is associated with important features of the last stages of oocyte maturation such as the resumption of meiotic maturation, inhibition of DNA replication between meiosis I and II and maintenance of the oocyte at metaphase II arrest until it is fertilized (Fauser *et al.*, 1999).

Very divergent views have been expressed about the role of various steroid hormones in the regulation of oocyte maturation. Observations are reported in this regard and both inhibitory and stimulatory effects of exogenous steroids, usually in high doses on isolated oocytes are observed (Guraya, 1985, 1997a, 2000a; Thibault *et al.*, 1987; Billig *et al.*, 1988; Jalabert *et al.*, 1991; Byskov *et al.*, 1995, 1999; Tllera *et al.*, 1997; Moor *et al.*, 1998; Baltsen and Byskov, 1999; Fauser *et al.*, 1999); inhibitors of steroidogenesis do not abolish the LH induction of meiosis in follicle-enclosed oocytes (Leibfried-Rutledge *et al.*, 1989) and thus it becomes doubtful whether the steroids play a major role in the control of meiotic arrest. How do such treatments induce fertilization abnormalities? General consensus emerged from these investigations show that oocyte requires specific intrafollicular steroid environment for complete maturation and that changes in this profile during ovum maturation result in gross abnormalities at fertilization. Such studies are required on the oocytes of women in different age groups to determine the degree of abnormalities in fertilization (Tevelde *et al.*, 2000; Bulletti *et al.*, 2001). However, Moudgal *et al.* (1996) have used aromatase, a specific inhibitor for determining whether there is a role for androgen in follicle/oocyte maturation, ovulation and preimplant embryo development. The results obtained suggest that although there is a clear requirement for androgen to support the reproductive cycle in the female, the need for androgen in regulating specific events is species dependent and thus studies in these lines are required to be carried out in human (Fauser *et al.*, 1999). Actually the importance of the steroid environment of oocytes during maturation becomes apparent after the consideration of various criteria such as fertilizability, that is sperm penetration and chromosomes decondensation or developmental ability (Jalabert *et al.* 1991; Fauser *et al.*, 1999; Wolf and Zelinski-Wooten, 2001; chapters 5 and 6). The results of these studies also indicate that more than one steroid may be needed for complete maturation of the oocyte. Byskov *et al.* (1995) have isolated and characterized meiosis-activating steroids from human FF and bull testes and synthesized the two closely related C 29 steroids. All these steroids induced resumption of meiosis in cultured cumulus enclosed oocytes and naked mouse oocytes showing their non-specificity across species and sex. Steroids directly or indirectly affect the proper development of the oocyte. In order to define the precise role (s) of steroid hormones in follicular and oocyte growth, it is essential to determine the nature and amount of steroids as well as the sequence and source of their production as discussed in detail by Guraya, (2000a) (see also Jalabert *et al.*, 1991; Fauser *et al.*, 1999). The FF ratio of progesterone to oestradiol is considered an important correlate of normal oocyte maturation. Determination of the intracellular molecular mechanisms controlling the shift between oestrogenic and progestational phases of follicular growth and maturation may provide useful relevant information (Guraya, 2000a). Various steroid hormones (e.g. testosterone, 17-oestradiol, and dihydrotestosterone) seem to function synergistically with the cAMP dependent factor under various conditions. The mechanisms and physiological significance of the steroid hormone synergism in the regulation of meiosis is required to be investigated at the molecular level (Leibfried-Rutledge *et al.*, 1989; Fauser *et al.*, 1999). Steroids may also affect cAMP dependant protein kinase (Guraya, 2000a). The synergism of steroids with factors that enhance cAMP consistently results in maintenance of arrest. Some kind of a fortuitous back- up mechanism appears to be provided by the follicular steroids. *In vitro* maturation of bovine oocytes is not influenced by the follicular diameter (Thibault *et al.*, 1976; Leibfried and First, 1979; Leibfried-Rutledge *et al.*, 1985) and possibly does not depend on the direct action of gonadotrophins (Guraya, 1997a, 2000a). However, addition of oestradiol and progesterone increases the proportion of bovine oocytes maturing *in vitro* in one study while these steroids do not have any effect (Thibault, *et al.* 1993). Thus, further investigations are required to determine more precisely the relationships among concentrations of FF steroids, oocyte maturation and fertilization in cattle as well as in human as some preliminary investigations show that in bovine follicle 5 mm and larger, high progesterone and low oestradiol levels in FF are correlated to the capability of oocytes to mature *in vitro* (Guraya, 2000a). Such studies are required to be carried out in women of variable age groups to determine the capability of their oocytes to mature *in vitro* (Tevelde *et al.*, 2000; Bulletti *et al.*, 2001; Wolf and Zelinski-Wooten, 2001). Virtually nothing is known about the intraovarian regulation of oocyte maturation in other ruminants including buffaloes. The porcine oocytes are, however, observed to produce factors that inhibit cumulus and mural GC steroidogenesis (Coskum *et al.*, 1995). The factor (s) seem not be transferred to somatic cells with gap junctions, and its effect is downstream of cAMP formation. However, Majumdar *et al.* (1988), after investigating maturation of slaughterhouse ovarian follicular oocytes of buffalo in culture and subsequent *in vitro* fertilization, stated that GVBD occurred in all the cases except the group where oestradiol-17b was used. The use of FF enhanced the GVBD percentage slightly. No significant increase in GVBD was seen with the addition of FSH-P. When the oocytes were incubated with epididymal spermatozoa in TC-199 or with addition of FSH-P, FF; 92 out of 360 oocytes fertilized; of these 57 were in 1-cell stage, 28 were in 2-cell, 6 were 3-cell and 1 was 4-cell stage. None of the oocytes of the oestradiol group fertilized. No beneficial effect on the addition of the hormone was found. Totey *et al.* (1991) have made studies of the effects of serum hormone and medium on *in vitro* maturation of buffalo oocytes. In the medium supplemented with 20% buffalo oestrus serum with additional FSH, LH and oestradiol-17b, 81% of the cultured oocytes showed a marked expansion of CCs, which was believed to form the specific feature of mature ova. Maturation rate was claimed to be significantly higher in both the mediums when 20% BES was added in combination with LH, FSH and oestradiol-17b supplemented with FCS. Hams-F-medium is claimed to show better maturation rates over TCM-199. The results of these studies are required to be extended and confirmed in different laboratories before these are accepted as the hormonal requirements for oocyte maturation in the buffaloes to be determined. Even the role of steroids in meiosis achievement of mammalian and human oocytes will continue to be controversial till antireceptors or specific antagonists are used (Thibault *et al.*, 1987). LH which is also secreted in higher concentrations in cattle and buffaloes, may function either by eliminating the inhibitory signal by blocking its transfer via the CCs oocyte junctions or by decreasing concentrations of sulfated glycosaaminoglycans in FF to ineffcetive concentrations by decreasing their synthetic rates and/or increasing the volume by terminating inhibitor production (Guraya, 1997a). Dekel *et al.* (1998) suggested that the GCs but not the TCs, mediate LH action to induce oocyte maturation in the rat. Chian *et al.* (2000) have investigated whether the rates of oocyte maturation, fertilization and development, as well as pregrancy rate, could be improved by I-KG primining 36 h before immature oocytes retrieval in patients with polycystic ovarian syndrome.

Further investigations need to be carried out to determine the complex molecular mechanism (s) involved in the regulation of meiotic maturation (Billig *et al.* 1988; Leibfried-Rutledge *et al.*, 1989; Eppig *et al.*, 1996; Guraya, 1997a, 2000a; Fauser *et al.*, 1999; Wolf and Zelinski-Wooten, 2001) not only in mammals but also in humans that some follicular component inhibits the expansion response of cumuli oophori until the inhibition is removed by preovulatory LH surge. A through knowledge of control mechanism(s) of ovum maturation at the molecular level will help us to improve the oocyte quality during follicle growth and ovulation in ruminants and humans especially women of variable age groups because primordial oocytes and ovarian stroma undergo aging changes in aged women (Chapter 1). It is well understood that the nature of follicular granulosa can influence the quality of an oocyte at ovulation. Therefore, there is always a concern about the use of gonadotrophins to induce follicular maturation and ovulation as the development of the oocytes could be distorted resulting in the production of embryos with limited potential for growth (Lauria and Gandolfi, 1992; Guraya, 2000a; Wolf and Zelinski-Wooten, 2001). This has been investigated in sheep and cattle by studying oocytes from ewes and heifers during natural cycle or following treatment with PMSG or FSH (Guraya, 1997a, 2000a). Such studies are required to be carried out on oocytes of women of variable age groups (Fauser *et al.*, 1999). The oocytes after their ovulation from the follicles are also in an unstable condition, as determined by their limited period of fertilizability, susceptible to aging, parthenogenetic activation, polyspermic fertilization and atresia (Wolf and Zelinski-Wooten, 2001). Such conditions can also be expected for human especially for the oocytes of aged women. Therefore, it is important to know how to overcome these problems and to find out more precisely which cellular and molecular events are responsible for the initiation of cellular alterations rather than merely responding to such change presents a great technical and analytical challenge of fundamental importance (Lauria and Gandolf, 1992; Wolf and Zelinski-Wooten, 2001). Therefore, sensitive micromethods need to be developed which will help us to make further precise analysis of physiological and biochemical and molecular processes involved in the growth and maturation of individual follicles and oocytes *in vivo* and *in vitro*. The results of such investigations will help to improve the oocyte quality after superovulation in farm animals, human and endangered species of mammals (Ware *et al.*, 1988).

3.3 TRANSCRIPTION AND TRANSLATION AND THEIR CONTROL

Numerous investigations are also being made of the qualitative and quantitative changes of RNA and protein synthesis and of metabolic alterations during ovum maturation both *in vivo* and *in vitro* (Guraya, 1985, 1997a, 2000a; Moor *et al.*, 1992; Thibault *et al.*, 1987; Moor and Gandolfi, 1987; Leibfried-Rutledge *et al.*, 1989; Wassarman and Albertini, 1994; Maric *et al.*, 1997; Shim-Chanseob *et al.*, 1997; Verrotti and Strickland 1997; Yang, 1997; Richard and Sirard, 1998; Moor *et al.*, 1998; Fauser *et al.*, 1999; Yen *et al.*, 1999; Wolf and Zelinski-Wooten, 2001). But it is yet to be determined more precisely whether production of RNA prior to GVBD is needed for the meiotic process or for later stages of embryocnic development. But Wabik-Sliz (1997) has suggested that the rate of meiotic maturation, sensitivity of oocyte investments to enzymes and deposition of granules in ooplasm are determined largely autonomously by genes acting in germ cells. However, Hue *et al.* (1997) have suggested that meiotic incompetance in goat oocytes is not due to absence of cyclin 21 protein as a potential pre-M-phase promoting factor subunit, but to a limiting amount of this protein. Mitotic cyclins (cyclins A and B) are clearly involved in meiosis and early embryonic cell cycles (Moor *et al.*, 1992; Taieb *et al.*, 1997; Fauser *et al.*, 1999). Richard and Sirard (1998) have observed that the 214 KDa protein secreted by theca cell monolayers plays a role in the process maintaining oocytes in meiotic arrest. The occurrence of abnormalities during fertilization appears to be due to incomplete maturation or absence of some cytoplasmic factors (Wassarman and Albertini, 1994; Fauser *et al.*, 1999; Wolf and Zelinski-Wooten, 2001). Both the block to polyspermy and the cytoplasmic factors needed for decondensation of sperm chromatin are not present in ovine oocytes fertilized during the GV or metaphase I stage (Moor and Gandolfi, 1987; Moor *et al.*, 1992). Synthesis of polypeptides having the ability to decondense sperm chromatin is seen between 12 and 18 hours after the induction of ovum maturation in sheep (Moor and Gandolfi, 1987; Moor *et al.*, 1992). Some proteins synthesized during ovum maturation persist till development (Wassarman and Albertini, 1994; Fauser *et al.*, 1999). The nature and relationship of protein synthesis to the maturation process of ova still remain to be defined more precisely (Moor and Gandolfi, 1987; Wassarman,

1988a; Leibfried-Rutledge *et al.*, 1989; Wassarman and Albertini, 1994; Fauser *et al.*, 1999; Wolf and Zelinski-Wooten, 2001) not only for different mammalian species but also for humans especially in women of variable age groups (Fauser *et al.*, 1999) because primordial oocytes undergo aging changes (Chapter 1). It is not yet known what influences are exerted by these aging changes during growth, maturation, fertilization and early development of oocytes (Tevelde *et al.*, 2000; Bulletti *et al.*, 2001). In this regard, it will be interesting to mention here the observations of Meng *et al.* (1996) who have observed that the protein synthesis in the maturing rat oocyte is regulated by cytoplasmic regulators rather than intrinsic nuclear components. No protein synthesis or transcription is needed for the GVBD in mouse oocytes (Wassarman and Albertini, 1994); none of the changes in protein synthesis accompanying meiotic maturation are due to mixing of the oocytes's nucleoplasm and cytoplasm. In contrast to this situation in the mouse, both trascription and synthesis of new proteins are needed for the GVD in the ungulate oocyte (Moor and Gandolfi, 1987; Moor *et al.*, 1992; Guraya, 2000a). Such a situation is required to be studied more precisely for human (Fauser *et al.*, 1999; Wolf and Zelinski-Wooten, 2001). However, all protein changes appear to occur subsequent to GVBD (Fauser *et al.*, 1999). However, a program of protein synthesis and expression is initiated at some stage between the signal to resume meiosis and GVBD. However, estimations of protein synthesis during preovulatory meiotic maturation in different species of mammals including mouse, rabbit, pig and sheep have shown that during this brief period of time, there occur major quantitative and qualitative changes in proteins (Leibfried-Rutledge *et al.*, 1989; Moor *et al.*, 1992; Wassarman and Albertini, 1994). Each stage of meiotic maturation is accompanied by specific qualitative and quantitative alterations in the pattern of protein synthesis (Fauser *et al.*, 1999). The control of stage-specific protein expression may be brought about by intrinsic program of messanger RNA activation of post-translational modification through phosphorylation and glycosylation (or both) during meiotic maturation and the post-fertilization period (Guraya, 1985, 1997a, 2000a; Leibfried-Rutledge *et al.*, 1989; Moor *et al.*, 1992; Wassarman and Albertini 1994; Miyano *et al.* 1996; Fauser *et al.*, 1999; Wolf and Zelinski-Wooten, 2001). Such studies are required to be carried out on the qualitative and quantitative changes in the pattern of protein synthesis as well as on their regulation in women of variable age groups (Tevelde *et al.*, 2000; Bulletti *et al.*, 2001). Two distinct developmental programs appear to direct the molecular changes during oogenesis as already discussed in Chapter 2 (Wassarman and Albertini, 1994; Guraya 2000a), whereas a separate maturation program controls reprogramming of the oocyte before ovulation. Various control mechanisms appear to influence protein synthesis during the maturation process and these molecular mechanisms involved in the continuity of reprogramming process are still to be determined precisely, not only in mammals but also in human, especially in women of variable age groups (Leibfried-Rutledge *et al.*, 1989; Moor *et al.*, 1992; Wassarman and Albertini, 1994). However, Sagata, (1996) has discussed the roles of Mos-MAPK pathway in oocyte meiosis and cellular transformation. Wassarman and Albertini (1994) have critically reviewed the previous literature about the expression of specific genes during oogenesis, which include C-mos, oct-30C-kit, mZP3, t-PA, LDH-B etc (Fauser *et al.*, 1999). But available evidence supports the suggestion that a program of 'schedule' of stage-specific protein expression starts during initial stages of preovulatory resumption of arrested meiosis in the oocyte (terminal events of oogenesis) and continues to function in the absence of germinal vesicle input through the early post-fertilization period (Leibfried-Rutledge *et al.*, 1988; Wassarman and Albertini 1994; Fauser *et al.*, 1999) as well also be discussed in Chapter 5. The evidence obtained so far suggests that changes in protein synthesis during meiotic maturation initiate at sometime between GV and GVBD stages and appears to be regulated at the post-transcriptional level (Leibfried-Rutledge *et al.*, 1989; Moor *et al.*, 1992; Wassarman and Albertini, 1994; Verrothi and Strickland, 1997; Fauser *et al.*, 1999). But the extent to which stage specific proteins specific of meiotic maturation derives from preformed cytoplasmic RNA templated and/or by post translational modification of precursor proteins is still to be determined more precisely. Meric *et al.* (1997) have discussed the potential role of nucleoplasmin in the remodeling of repressive ribonucleoprotein particles containing maternal mRNA to facilitate translational activation. Nussbanm and Prather (1995) have shown the differential effects of protein synthesis inhibitors on porcine oocyte activation.

Actually, the entry of the spermatozoon starts an early embryonic program, which continues to persist until maternal regulation is terminated and development becomes directed by the embryonic genome (Howlett and Bolton, 1985; van Blerkom, 1985; Wassarman and Albertini, 1994; Fauser *et al.*, 1999; Wolf and Zelinski-Wooten, 2001 chapters 5 and 6). Dominance of transcription as the major determinant of gene expression does not take place until embryonic control of development begins. The results of various investigations have shown that completion of meiosis, processing of the sperm nucleus following fertilization, resumption of meiotic cycles and early embryonic development are directed mainly by maternally-derived informational macromolecules (RNA and protein) formed during oogenesis (Davidson, 1986; Fauser *et al.*, 1999; Wolf and Zelinski-Wooten, 2001 Chapter 2 as well be discussed in Chapter 6). Depending upon the species, the synthesis of transcripts and proteins may not be completed until the periovulatory period and meiotic maturation is also complete. These are stored for use during fertilization and early development (Wassarman and Albertini, 1994; Fauser *et al.*, 1999; chapters 5 and 6). Depletion of glutathione during bovine oocyte maturation reversibly blocks the decondensation of the male pronucleus and pronuclear apposition during fertilization (Sutovaky and Schatten, 1997).

Further investigations are required to be carried out to determine more precisely patterns of transcription and translation and their regulation during ovum maturation in different species of mammals (Moor *et al.*, 1992; Wassarman and Albertini, 1994) and especially for women in variable age groups (Fauser *et al.*, 1999; Tevelde *et al.*, 2000; Bulletti *et al.*, 2001; Wolf and Zelinski-Wooten, 2001) because the aging process of primordial oocytes and stroma may cause some changes in the patterns of transcription and translation during oocyte growth (Chapter 2) and maturation as well as during early embryogenesis after fertilization as will also be pointed out in Chapter 6. Regulatory proteins involved in stage-specific transcription and translation, as well as in regulation of meiosis, are apparently produced during oocyte growth (Chapter 2). But they remain to be identified and characterized in future investigations by using recombinant DNA and immunological probes (Wolf and Zelinski-Wooten, 2001). Moor *et al.*, (1985) observed that all the oocytes from untreated sheep, which are incubated in follicles without hormonal addition, show a dictyate nucleus. In comparison, the addition of hormones in the culture medium causes the GVBD in almost 8 of the oocytes (Moor *et al.*, 1992). The proteins synthesized by the oocyte obtained from unstimulated follicles in culture are undistinguishable from those in oocytes taken directly from follicles *in vivo*, showing that the culture method does not cause any alterations in the oocyte. Culture media for mouse oocyte maturation affect subsequent development (van de

Sandt *et al.*, 1990). Wallace and Rajamahendran (1993) have studied the effects of zeralenone on *in vitro* early development of mouse embryos and maturation of bovine oocytes. Maturation rate for control oocytes was 74%, which was not different from that of other treatment groups. The administration of FSH into the sheep does not impair the growth of follicles and 70% of oocytes taken from follicles exposed to hormones in vitro posses a full programming of protein synthesis similar to those maturing in untreated animals (Moor *et al.*, 1992). In contrast, an injection of PMSG into the mother appears to have deleterious effects on the pattern of synthesis in many oocytes. More than one quarter of the oocytes shows variations of protein synthesis and it is possible that many of these have already been activated before the follicles are excised. Continued studies on the regulation of transcription and translation process during oocyte growth and maturation, fertilization and early embryogenesis are recommended to be carried out *in vivo* and *in vitro* systems under normal physiological and pathological situations (Wolf and Zelinski-Wooten, 2001). The results of such studies at the cellular and molecular levels will be very rewarding to determine the causes of abnormalities or disorders during embryogenesis, which are now of common occurrence not only in mammals but also in humans (Fauser *et al.*, 1999; Wolf and Zelinski-Wooten, 2001); the causative factors for these developmental abnormalities are still to be determined (Chapter 7).

4

Ovulation

The term ovulation is used here to describe the processes, which lead to the release of the mature oocyte from the follicle. The release of the mature oocytes is accompanied by various cellular and molecular changes in the follicle wall, which will be described here. As a result of a thinning in portions of the follicle wall, the rupture or hole is formed in the follicle wall through which the oocyte is released (Figs. 1, 16 and 17). Various studies indicate that the release of the oocyte is an active process involving contractions of smooth muscles in the follicle wall. Actually various biochemical regulators involved in the process of ovulation at the follicle level have been determined. The chemical mediators appear to arise and function within the follicle wall or ovary in the process of ovulation (Karam *et al.*, 1999). The regulation of many of the ovarian mediators will be described in relation to gonadotrophins such as FSH, LH, hCG and PMSG which are being used extensively to investigate the regulatory factors involved in the mechanism of ovulation *in vivo* and *in vitro* (Findlay *et al.*, 1996; Kebly *et al.*, 1999; Filicori *et al.*, 1999; Sheffield *et al.*, 2000). The use of such gonadotrophins may cause superovulation, depending upon the gonadotrophins and its dose, time of administration during the oestrous (or menstrual in human) cycle of mammalian species including humans are discussed in the previous book and in a recent papers (Lenton *et al.*, 2000; Felberbaumn *et al.*, 2000; Guraya, 2000a) and will not be described. It is important to reduce the multiple gestation pregnancy rate in assisted reproductive treatment programmes, despite the pressure from some patients to transfer more embryos in order to improve success. Here emphasis will be laid on the cellular and molecular aspects of mechanism of ovulation. It is well established now that at or around its maximum peripheral concentration, oestradiol triggers the ovulation inducing discharge of LH from the anterior pituitary (Guraya, 2000a). In order to define the factors those regulate mechanism of ovulation in human, various morphological, histochemical, biochemical and physiological changes occurring in granulosa, basal lamina, theca interna, surrounding stroma including its blood supply, and surface epithelium will be discussed. For the sake of clarity and brevity, extensive literature will not be cited and only specific points of view with focus on issue, which are the subject of some controversy will be emphasized at the cellular and molecular level, for more details a reference is made to Guraya (2000a). All these alterations in the follicle wall during ovulation occur in response to preovulatory LH surge, which is well known to cause ovulation in human and other mammalian species (Guraya,

1974, 1985, 1997a, 2000a). Actually preovulatory surge of gonadotrophins causes a series of orderly sequenced alterations in the mature preovulatory ovarian follicle which include resumption of oocyte maturation and germinal vesicle break down (discussed in Chapter 3), initiation of the luteinization of the granulosa cells (Guraya, 1971a, 2000a) and restructuring of the follicle wall with resultant rupture of follicle wall and release of a fertilizable ovum or egg.

4.1 MECHANISM

There is a general agreement that the first macroscopic indication that ovulation is about occur is the appearance of the macula pellucida or stigmata (Figs. 1, 16 and 17) (Guraya, 1971b, 1974, 1985, 1997a, 2000a). The size and form of the stigmata vary greatly in different species of mammals and submammalian vertebrates (see Guraya, 1985, 1986, 1989, 1997, 2000a). Careful studies of ovulatory stigmata in various species have shown that these structures are avascular. There occur alterations in the blood vascularity of stigmata during ovulation. On the bases of morphological observations on the process of ovulation made by various workers it can be concluded that ovulation generally does not occur in an explosive manner but is a steady, continuous, and slow process, thus negating the role of intrafollicular pressures advocated by some workers (Guraya, 1974, 1985, 1986, 1997a, 2000a). On the basis of formation of follicular epithelial infolds due to shrinkage of the follicle during the preovulatory period, the role of intrafollicular pressure in the process of ovulation has also been discussed (Guraya, 1974, 1999a, 2000a). The smooth muscle-like cells demonstrated recently with electron microscopy in the theca externa of the follicle may be concerned with the preovulatory and postovulatory shrinkage of the follicle but not with the increase in intrafollicular pressure (Guraya, 1974). During ovulation, the theca and granulosa layers separate as a result of dissolution of the basement lamina, permitting blood vessels to invade the granulosa cells which simultaneously start luteinizing (i.e., the characteristics of steroid gland cells are developed by the involment of cAMP and form the major source of progesterone in the corpus luteum (Guraya, 1971a, 1997a, 2000a), theca interna cells are also stimulated by LH as evidenced by the increased production of progesterone, androgens, postaglandins and plasminogen activator. Colgin and Murdoch (1997) have suggested that secretion of urakinase-type plasminogen activator by ovarian surface epithelium and consequent up-regulation within the neighbouring tunica albuginea and follicle theca forms a contributing factor in the mechanism of ovulation.

On the basis of cellular biochemical and biophysical changes, which occur in various components, different theories have been put forth to explain the mechanism of ovulation such us (1) intrafollicular pressure as already pointed out, (2) contraction of smooth muscle cells, (3) nervous control, (4) a neuromuscular mechanism, (5) vascular changes and (6) enzymatic digestion (Guraya, 2000a). At present no single theory is accepted as the sole cause of ovulation; rather a combination of these phenomena is believed to be involved in the ovulation as in recent years reviewed in detail by several workers (Guraya and Dhanju, 1992; Espey and Lipner, 1994; Guraya, 1997a, 2000a). However, various contributions made in support of these theories have enhanced our knowledge of the cellular and subcellular mechanisms associated with steroidal and nonsteroidal biosynthesis, collagenolysis and vascular alterations related to ovulation but have not yet produced the precise evidences for the follicular rupture. The edema and inflammation are believed to be the determinants, of ovulation and instead synthetic mechanisms initiated in the preovulatory follicle in response to LH surge play a crucial role in ovulation (Guraya and Dhanju, 1992). This suggestion is supported by the fact that the cascade of various changes associated with ovulation can be interrupted by blockade of protein synthesis by cycloheximide or actinomycin D, by inhibition of steroid biosynthesis with cyanoketone or aminoglutethemide, antiprogesterone antiserum and antitestosterone antiserum by inhibition of prostaglandin synthesis with indomethacin and by interruption of the plasminogen activator-plasminogen cascade (Guraya and Dhanju, 1992; Espey and Lipner, 1994; Guraya, 1997a, 2000a; Fauser *et al.*, 1999). Fujwara *et al.* (2000) have discussed expression activity and regulatory mechanism of amino peptidases in ovarian follicle development and the effect and use of peptidase inhibitor, bestatin, in induction of ovulation. An amino peptidase inhibitor bestatin, is observed to enhance gonadotrophin-stimulated ovulation (Nakamura *et al.*, 1996). The results obtained indicate that membrane-bound peptidase(s) present on murine ovarian cells forms an important regulating factors(s) of follicle growth and/or ovulation. Such studies are required to be carried out for human species in young and aged women (Tevelde *et al.*, 2000; Bulletti *et al.*, 2001).

Now it is well established that prior to ovulation, various morphological and chemical alterations cause a progressive dissociation and decomposition of various cellular layers surrounding the stigmata (Fig. 16), which have been attributed to the effects of activities of various proteolytic enzymes especially plasmin and collagenase activities (Guraya and Dhanju, 1992, Eopey and Lipner, 1994; Guraya, 1997a, 2000a). The changes in lysosome enzyme activities in preovulatory follicles are indicative of alterations in the process of tissue, remodeling required for ovulation of mature and normal oocytes (Banos *et al.*, 1996). This has been strongly supported by the observation that the effect of the intrafollicular injection of several enzymes on the rupture of the rabbit Graafian follicle (Guraya, 1974, Guraya and Dhanju, 1992; Espey and Lipner, 1994); other studies have revealed an increase in the amount of collagenolytic enzymes in the follicle fluid during the expansion period. On the basis of their evidence showing similarities between enzymically induced rupture of the follicle and coitally induced ovulation, Espey and co-workers postulated that the structural changes preceding follicular rupture might be the result of proteolysis of the follicle wall (Espey and Lipner, 1994). Furthermore, they suggested that if proteolysis were indeed responsible, stigma formation would follow as result of generalized changes in the follicular wall. Electron microscope studies have shown that there is disintegration of the connective tissue elements in the wall of the Graafian follicle before rupture occurs. Byskov (1969) observed that the ovulation gap in the preovulatory follicle of the mouse ovary formed by successive degeneration of the cell layers in the apex, except the granulosa layer, starting at the outermost layer, the epithelium; these changes are also closely accompanied by structural alterations in the walls of blood capillaries. Guraya (1971b) observed that as a result of edema in the theca externa and the surrounding stroma of preovulatory follicles in rabbit, the cells become increasingly separated from each other as extracellular fluid accumulates. Simultaneously, the physical and chemical characteristics of the nucleic acids (RNA and DNA) of the cells of surrounding stromata and stigmata are apparently altered during ovulation, as positive reactions are much reduced in these cells and disappear completely from cells in the area of the stigma (Fig. 16). Similar morphological and histochemical changes have also been observed in the surrounding stroma of newly ruptured follicles of the human (Guraya, 1974), which can be presumed to undergo disintegration during ovulation. On the basis of morphological and histochemical changes in the surrounding

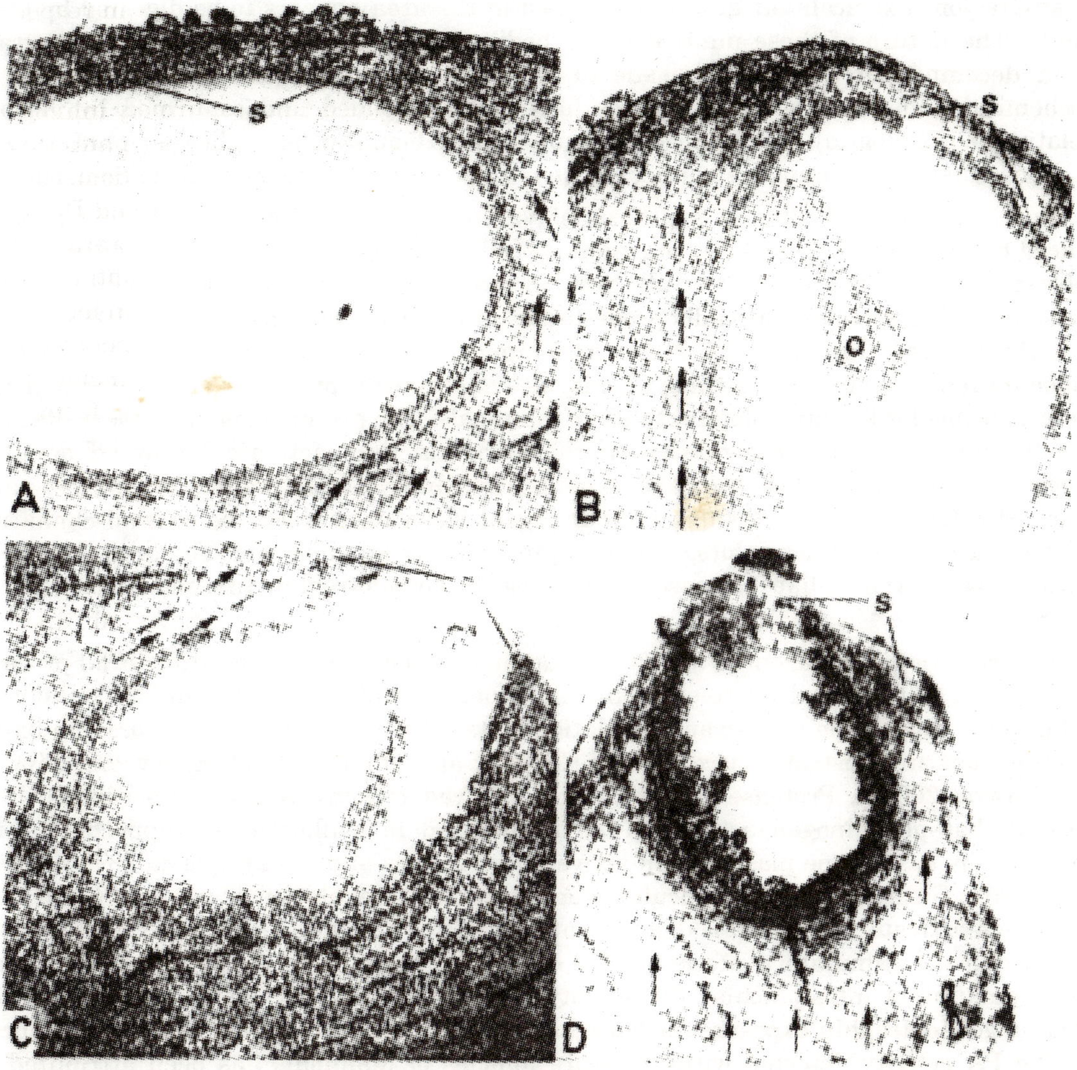

Fig. 16

ovarian stroma associated with preovulatory swelling and ovulation, Guraya (1971b) suggested that the rupture of the follicle may be due to catabolizing factors, and that it may be useful to study the nature, concentration, regulation and cellular sites of synthesis of various enzymes that cause changes in various components of preovulatory follicle in the ovaries of young and aged women these may be lysosomal hydrolyses, for example, DNAase, RNAase, cathepsins (*i.e.*, proteolytic enzymes), which may be set free to bring about the changes in nucleic acids. These results of histochemical study are strongly supported by the electron microscopic observations (Espey and Lipner, 1994), who found that the fibroblasts in the tunica albuginea and theca externa of all stages of ovulatory follicles contain microvesicles which cluster together and protrude from the surface of the plasma membrane usually develop in the vicinity of free ribosomes or granular endoplasmic reticulum, and protrude from any region. Extracellular ground substance in the area appears to be digested by their contents. The nature of these multivesicular bodies suggests that they have an important part in decomposing connective tissue of ovulating Graafian follicles. The results of histochemical and ultrastructural studies have been extended and confirmed by various correlative morphological, biochemical, immunological, biophysical studies on preovulatory follicles of different mammalian species *in vivo* and *in vitro* systems as discussed here.

The regulation and sites of synthesis of various enzymes suggest that they may be produced by the surface epithelium or granulosa cells including cumulus cells or surrounding fibroblasts including thecal layer, or all three, in the region of stigmata, possibly in response to gonadotrophins, steroids especially progesterone, and prostaglandins (Guraya, 1985, 1997a, 2000a; Dhanju and Guraya, 1992; Espey and Lipner, 1994). Three types of proteases such as plasminogen activators, collagenolytic enzymes and neutral proteases are now considered to be responsible for the proteolytic activity in mammals. The plasminogen activators-plasmin and collagenase may be interrelated in that plasmin may activate latent collagenase producing, a cascade effect at the time of ovulation (Guraya and Dhanju, 1992; Espey and Lipner, 1994; Guraya, 1997a, 2000a). Very divergent views exist about the nature, regulation and sites of synthesis of activators, inhibitors, modulators etc. involved. The granulosa cells produce a noval trypsin-like protease in response to gonadotrophin treatment (Blacsak *et al.*, 1989).

The regulation of collagenase activity in mammals appears to be a multifactorial process, involving both cellular and micro-environmental factors, which include cellular modulation of collagenase, proenzyme release and regulation of its activation by local activators, inhibitors and alterations in substrate susceptibility (Guraya and Dhanju, 1992; Espey and Lipner, 1994; Guraya, 2000a). Proteases appear to be involved in activation of collagenases. It is suggested that plasminogen activator (PA) is involved in ovulation, as supported by its suppression by tissue-type plasminogen activator antibodies and a2- antiplasmin (Espey and Lipner, 1994; Guraya, 2000a). Follicular stage dependent regulation of rat granulosa cells plasminogen activator by TGF-a is demonstrated *in vitro* (Karakji and Tsang, 1995). IGFBP-3 by blocking the stimulatory effects of hCG on the ovulatory process appears to contribute to the regulation of intrafollicular PA activity during follicular development and ovulation evoked by gonadotrophin exposure, at least in part, via neutralizing endogenously produced IGF. The PA activity detected within mature follicles of mammals has been attributed to two different activators, urokinase plasminogen activator (UTP) and tissue-type plasminogen activator (tpa), which are stimulated by the gonadotrophin surge (Hsueh *et al.*, 1989) possibly in response to steroid hormones (Guraya and Dhanju, 1992). A positive relationship is observed between steroids and the ovarian plasmin activity during ovulation. PA and plasmin are involved primarily in early alterations resulting in follicle rupture i.e., collagenase activation, which is blocked by various inhibitors of collagenase activity (Espey and Lipner, 1994; Guraya, 2000a). Plasmin activates latent ovarian collagenase precursor (Guraya and Dhanju, 1992; Espey and Lipner, 1994). Collagenolytic activity increases in the preovulatory follicle several hours after the LH surge and attains its maximum just before ovulation occurs, suggesting that collagen-degrading enzymes may be involved in the mechanism of ovulation as also supported by its blockage with a specific inhibitor of mammalian collagenase. In rat pro oestrous follicles, both LH and FSH are equally effective in stimulating PA activity (Guraya and Dhanju, 1992; Espey and Lipner, 1994; Guraya, 2000a). But the precise roles of these gonadotrophins in stimulating PA activity in the ovaries of young and aged women are required to be determined at the molecular levels (Tevelde *et al.*, 2000; Bulletti *et al.*, 2001). However, the gonadotrophins induce the transcription of tPA mRNA suggesting their role at the transcriptional level to regulate tPA activity. Both the peptide hormone, relaxin and LH-RH also increase tPA activity in the rat granulosa cells but the mechanism remains to be determined tPA is now known to be involved in ovulation as blockage of ovulation and subsequent cyst formation results from inadequate tPA activity in manipulated follicles of gilts (Whisnant *et al.*, 1998). Such studies at the molecular level are required to be carried out in women of variable age groups. Jia *et al.* (1990) have observed the synergistic effect of gluco-corticoids and androgen on the hormonal induction of tPA activity and mRNA concentrations in granulosa cells. Beta-adrenergic agents can stimulate tPA and mRNA levels in cultured granulosa cells (Oikawa and Hsueh, 1989). Liu-Yixin *et al.* (1998) have reported the prolactin regulation of tPA and PA inhibitor type-1-gene-expression in eCG-primed rat granulosa cells in culture. Epidermal growth factor also shows similar effects (Galway *et al.*, 1989). Tissue-type PA is observed within the denuded rat oocyte, suggesting the possibility of its association with meiotic maturation (Goetz *et al.*, 1991). These authors have also suggested that the prostaglandins apparently function to dissociate the follicle wall by inhibiting the formation of new collagen fibres rather than by stimulating their degradation. The PA-plasminogen system causes controlled proteolysis and tissue degradation during the rupture of the follicle wall at the time of ovulation. The increased production of PA is known to convert the plasminogen in the follicular fluid and extracellular fluid to plasmin, which acts on latent collagen attached to the collagen fibres. This induced collagenolysis and serine proteases then cause the complete proteolysis of the collagen. The net result is to decrease the tensile strength of the follicle wall to the point at which rupture occurs under the existing intrafollicular pressure of 15 to 20 mm (Guraya, 1985, 1997a, 2000a; Espey and Lipner, 1994). Neutral protease activity present within the rat follicle is greatest during the oestrous phase and is suggested to be regulated by LH (Kaur and Guraya, 1986; Guraya and Dhanju, 1992). Neutral protease activity is not only stimulated by LH but also by FSH, thyroid-stimulating hormone, and growth hormone, suggesting an important role in mammalian ovulation (Kaur and Guraya, 1986; Guraya and Dhanju, 1992). Dhanju *et al.* (1990) have demonstrated a stimulatory effect of progesterone and oestradiol- 17b on ovulation and ovarian neutral proteinases which may be affected either directly through protein synthetic activity or through some other factors involved in ovulation (Dhanju *et al.*, 1991a, b; Guraya and Dhanju, 1992). The role of tyrosine kinase in gonadotrophin induced ovulation in the rat ovary is suggested (Shimamoto *et al.*, 1998).

Various studies carried out on the role of enzymes and other chemical factors to explain the mechanism of ovulation in the rodents especially rat as discussed above are required to be carried out on the follicle wall during ovulation in human (*i.e.*, in women of variable age groups) (Tevelde *et al.*, 2000; Bulletti *et al.*, 2001). However, an early response of the preovulatory follicle to LH is increased steroid biosynthesis, and blockade of steroid synthesis inhibits ovulation, suggesting that steroid cascade is involved in the production of PA as supported by studies of Dhanju *et al.* (1990, 1991, 1992) who have found a positive relationship between steroids and ovarian neutral protease and PA activities. The prostaglandins also increase PA production, as does steroid (progesterone). Although after the LH surge proteolytic activities at the apex are mainly involved in follicle rupture but this does not cause the complete destruction of the follicle wall. This process only permits the cellular and vascular remodelling, resulting in the formation of the corpus luteum. Complete disintegration is only confined to the apex of the follicle and here other factors are believed to be involved. Our knowledge is still meagre about the molecular mechanisms involved both in the increase of blood flow during the LH surge and later in the blood stasis at the follicular apex (reviewed by Thibault and Levasseur, 1988; Guraya and Dhanju, 1992; Guraya, 2000a).

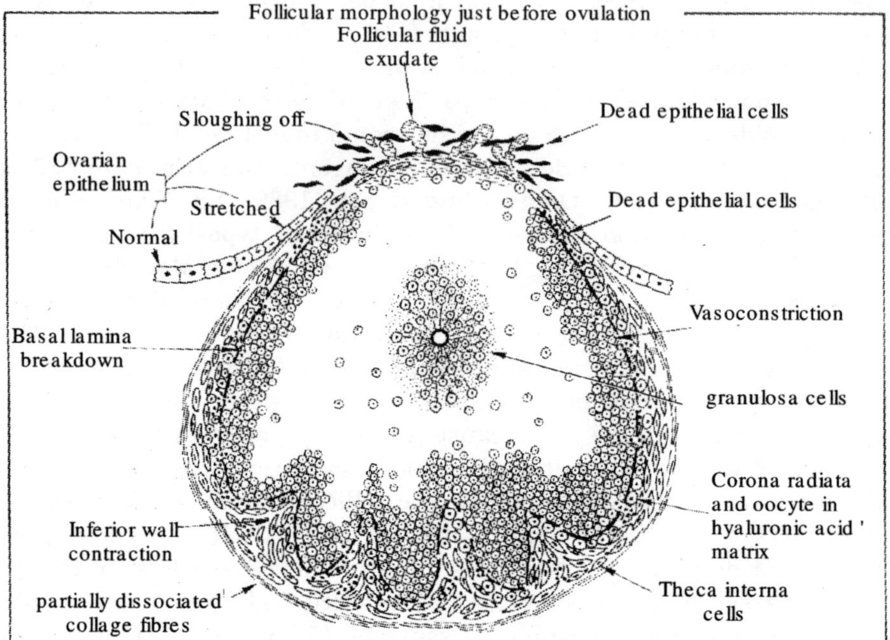

This figure presents the histological and morphological changes occurring in the preovulatory follicle and enabling follicular rupture and oocyte expulsion. (Redrawn from Thibault and Levasseur, Human Reprod., 1988, 3, 513-523).

Fig. 17

The interaction between histamine, prostaglandins and catecholamines is considered to be essential for ovulation (Thibault and Levasseur, 1988; Espey and Lipner, 1994; Guraya, 2000a). Prostaglandin G/H synthase-2 can be used as a marker for follicular commitment to ovulation during ovarian hyperstimulation protocols (Liu, Jianmin *et al.*, 1998). But it is still not known how they precisely contribute to the apical disintegration of the follicle wall. Because of the involvement of prostaglandins and histamine, the ovulation is linked to an inflammatory response (Guraya, 2000a). Aspirm and naprozen are found to reduce significantly ovulatory efficiency and prostaglandin production both *in vivo* and *in vitro* in hCG-treated rabbits (Zanagnalo *et al.*, 1996). Although involvement of prostaglandins in ovulation is well established but their individual roles (i.e., stimulation or inhibition of ovulation) and sites of their production remain to be determined more precisely (Guraya and Dhanju, 1992; Espey and Lipner, 1994; Guraya, 2000a). They do not activate plasminogen or collagenase, which are found to increase after the LH surge or administration of hCG in females in which ovulation is inhibited by indomethacin (Guraya, 2000a). Indomethacin is well known to block ovulation *in vivo* and *in vitro* in various mammalian species including cows and ewes (Guraya and Dhanju, 1992; Guraya, 2000a). But such studies remain to be carried out in human.

The synthesis of prostaglandins is increased by the direct action of gonadotrophins (FSH, LH, hCG, hMG and PMSG) mediated by cAMP and not by steroids (Espey and Lipner, 1994; Guraya and Dhanju, 1992; Guraya, 2000a). However, progesterone can activate PGE2-9-keto reductase resulting in an inversion of $PGE_2/PGE_2\alpha$ ratio that is required for ovulation in the sheep (Murdoch *et al.*, 1986). The progressively delayed expression of prostaglandin G/H synthase-2 in species with longer ovulatory processes such as in equicine preovulatory follicles supports its role as a molecular determinant of the species-specific length of the ovulatory process (Serios and Dore, 1997). Steroids are believed to contribute to the regulation of prostaglandin production, which in turn may regulate enzyme production (Espey and Lipner, 1994; Guraya, 2000a). A gonadotrophin (LH) induced preovulatory increase in follicular prostaglandin, produced by both theca and granulosa cells (Guraya, 2000a) is required for ovulation. The concentrations of prostaglands in the follicular fluid and the sites of their synthesis have been investigated in larger mammals including pig, sheep, goat, cow etc. (see Goetz *et al.*, 1991 for references) but not in human.

The biochemical and hormonal mechanisms regulating the synthesis of prostaglandins in follicles at a specific stage of development and a specific time after the LH surge continue to be unclear, although indirect evidence has suggested an important role for PG synthase (Guraya and Dhanju, 1992; Espey and Lipner, 1994; Guraya, 2000a). LH regulates ovarian follicular prostaglandin production by two different, but complementary modes of action-an increase in archidonic acid availability and an increase in prostaglandin synthatase-both of which depend on the follicular cell type and stage of differentiation, may be functioning (Espey and Lipner, 1994; Guraya, 2000a). Goetz *et al.* (1991) have discussed the molecular aspects of regulation of synthesis of PG synthase. The protein kinase C is believed to be involved in the synthesis of prostaglandins. Protein synthesis is necessary for phorbol ester (PE), stimulation, and PE appears to increase the production of prostaglandins and the availability of fatty acid precursor for prostaglandin synthesis. The rapid induction of prostaglandin endoperoxide synthase in rat preovulatory follicles by LH and cAMP can be blocked by inhibitors of transcription and translation (Wong *et al.*, 1989). Kaur and Guraya (1989) have observed that on indomethacin treatment, the activity of ovarian neutral proteinases and 3β-HSDH is not affected but the total proteins are decreased significantly in the ovaries of rat, suggesting that prostaglandin inhibition caused by indomethacin possibly blocks ovulation by interfering with protein metabolism (reviewed by Guraya and Dhanju, 1992). The prostaglands are required for the synthesis of some specific proteins involved in the ovulatory process and this suggestion remains to be extended and confirmed by further cellular and molecular studies.

The role of arachidonic acid metabolites in follicular rupture at ovulation is reported (Espey and Lipner, 1994; Guraya, 2000a). The arachidonic acid metabolites appear to play an essential role in gonadotrophin activation of ovarian collagenolysis, needed for follicle rupture at ovulation. This is supported by the fact that inhibitors of cyclooxygenase and lipoxygenase pathways of arachidonic acid metabolism inhibit follicle rupture at ovulation (Goetz *et al.*, 1991; Guraya and Dhanju, 1992; Espey and Lipner, 1994; Guraya, 2000a). Both cyclooxygenase and lipoxygenase do not show any effect on follicular PA, but inhibit LH-induced increase in ovarian collagenolytic activity. The clear effects of arachidonic acid metabolites on follicular rupture and their production in the ovary justify inclusion of eicosanoids among local ovarian hormones. But further studies need to be carried out to localize ovarian receptors of eicosanoids as well as to define the cellular and molecular mechanisms underlying their involvement in follicular rupture and corpus luteum function.

The prostaglandins show numerous and varied actions, among which are smooth-muscle contractile ($PGE_2\alpha$) or relaxant (PGE_2) activity, activation of adenylate cyclase participation in the inflammatory reaction and activation of proteolytic enzymes (Goetz *et al.*, 1991; Guraya and Dhanju, 1992; Espey and Lipner, 1994; Guraya, 2000a). Although the prostaglandins affect the smooth muscle of the ovary and possibly follicle, a relationship to ovulation is still not established. One of the important actions of the leukotrienes is in their chemotactic effect on polymorphoneutrophils and their involvement in the inflammatory reaction. Various studies have clearly demonstrated the involvement of prostaglandins and leukotrienes in follicular rupture as discussed in detail by Goetz *et al.*, (1991). But the molecular mechanisms of action of prostaglandins during ovulation remain to be defined. Harwitz *et al.*, (1997) have studied *in vitro* modulation of PA activity, prostaglandin E and nitric oxide production by interleukin-I (IEF) in PMSG-primed theca interna cells. The results obtained attest to a pleiotropic response of PMSG-primed theca interna cells to IEF, and suggest a paracrine autocrine function of theca interstitial cells compartment in ovulation and, corpus luteum formation (Guraya, 2000a). According to Machelon *et al.* (1995) both granulosa cells and macrophages are actively involved in the ovarian production of IL-I-beta at ovulation and the ability of granulosa cells to produce IL-I-beta may require macrophages which are now implicated in folliculogenesis and ovulation (Araki *et al.* 1996; Fauser *et al.*, 1999). The human preovulatory follicle forms the source of chemotatic IL-B which plays a role in cyclic ovarian events, such as ovulation (Runesson *et al.*, 1996; Aricu *et al.*, 1996) as also suggested for the rat (Hoek *et al.*, 1998) and rabbit (Uljaka *et al.*, 1998). Murdoch *et al.*, (1997) have suggested that a bioactive soluble form of tumor necroses factor alpha generated from endothelial cells of preovulatory ovine follicles has an obligatory role in the mechanics of apical follicle weakening and ovarian rupture. The inhibition of nitric oxides (NO) has suggested that its pathway may play an important role in the regulation of ovulation and the mediation of IL-IB preovulatory effects (Bonello *et al.*, 1996). These are believed to be primarily vascular effects, but also a non-vascular component, to the NO regulation of ovulation, with both components indirectly effecting ovulatory leukocyte distribution and steroid secretion. Yamauchi *et al.* (1997) have, however, suggested that NO induces follicle rupture in rabbit ovaries at least partly by stimulating prostaglandin production.

The prostaglandins and histamine show independent actions as indomethacin does not inhibit the induction of ovulation by histamine, but antihistamines can prevent follicular rupture *in vivo* and *in vitro* (Goetz *et al.*, 1991). The decrease in the number of mast cells forming the source of histamine in the wall of bovine dominant follicle is related to ovulation as supported by histamine of the sympathetic nerve-mediated contractions in the bovine ovarian follicle wall *in vitro* (Guraya, 1997a, 2000a). The histamine induces contraction via H_1 receptors and relaxation via the H_2 receptors in a situation analogous to that observed with adrenergic receptors. The general concensus is that histamine released from mast cells and the theco-adrenergic agonist effects increase the hyperemia by affecting the contractility of the endothelial cells, the pericytes and the postcapillary venules (Goetz *et al.*, 1991; Espey and Lipner, 1994). Recent molecular studies on the expression of several members of neurotrophin family, including nerve growth in the mammalian ovary also support the role of innervation in ovulation (Dissen *et al.*, 1996). The various sites of action of histamine are suggested to include edema, and loosening of the follicle wall, vascular resistance, contractile effects, and modulation of neurotransmitter and steroid metabolism. Various lines of evidence indicate that, while not essential for ovulation, histamine may perform a facilitatory role along with other factors in ovulation. The other vasoactive substances that may play a role in ovulation include provenin, renin, angiotensin II; kinins, kullikretin activity etc. (Goetz *et al.*, 1991). These authors have discussed the possible hormonal regulation of formation sites of synthesis and storage as well as the mechanism of action of various vasoactive substances during ovulation. Feral *et al.* (1996) have shown the presence of high affinity angiotensin II receptor type I on granulosa and thecal cells of rabbit preovulatory follicles, contrasting with previous observations showing its presence on granulosa or theca cells. Such studies are required to be carried out at the molecular level in human (Fauser *et al.*, 1999).

Various lines of evidences have indicated that neural innervation of the follicle wall (or neurotransmitters) possibly modulates the mechanism of ovulation and that the smooth muscle in the follicle wall is suggested to play either no critical role in ovulation, or a major role (reviewed by Goetz *et al.*, 1991; Espey and Lipner, 1994; Guraya, 1997a, 2000a). The evidence for a role of innervation in ovulation is produced from three sources (1) presence of adrenergic and cholinergic innervation in the ovary especially in the follicular wall (Guraya *et al.*, 1991; Guraya and Dhanju, 1992) (2) effects of pharmacological agents (adrenergic agonists and antagonists on stimulation or inhibition of ovulation (Goetz *et al.* 1991); and (3) studies on contractions caused by neurotransmitters in the follicular wall (Goetz *et al.*, 1991). The results of studies in these lines in the human will be very rewarding for induction of ovulation in those women who do not respond to exogenous gonadotrophins (Wolf and Zelinski-Wooten, 2001). However, inspite of interspecies variation in the density of innervation there occur similar distributional patterns for the ovarian nerves which innervate the stroma, vasculature and the follicular wall (Guraya *et al.* 1991; Guraya and Dhanju, 1992); adrenergic nerves are more abundant than cholinergic nerves. Innervation of the vasculature varies to a greater extent than does that of the stroma or follicular wall. Bovine and ovine follicles themselves show more dense innervation than that of the stromal tissue. The close contacts occur between the smooth muscle cells in the theca externa and the nerve terminals. Contractions of follicle wall strips *in vitro* by the production of catecholamines mediated either by tyramine or electrically strongly suggests an important role for adrenergic agents in follicular rupture-oocyte expansion (Goetz *et al.*, 1991). However, the conclusion at present is that smooth muscle cells contraction in the follicle wall during ovulation is involved in the detachment of the cumulus oophorus, expulsion of the follicular contents after opening of the apical wall, related vascular phenomena and collapse of the follicle and its transformation into a corpus luteum (Guraya, 1985, 1997a, 2000a; Thibault and Levasseur, 1987; Guraya and Dhanju, 1992; Espey and Lipner, 1994). Angiotensin II is found to induce ovulation and oocyte maturation and stimulates the production of oestrogen and prostaglandin by perfused rabbit ovary *in vitro* via the AT_2 receptors (Yoshimura *et al.*, 1996; Nakamura *et al.*, 1996; Kuji, *et al.*, 1996). Thus, locally produced Ang II may be part of a noval intraovarian paracrine or autocrine control mechanism during the ovulatory process. The results obtained by Kuji *et al.*, (1996) also support the involment of Ang. II in the process of ovulation or fecundation.

Catecholamines showing an increase in their levels with the approach of ovulation regulate the contraction of follicles and ovaries demonstrated with the use of pharmacological agents and their action in this regard is very complex (Goetz *et al.*, 1991; Guraya and Dhanju, 1992; Espey and Lipner, 1994; Guraya, 2000a); α-receptors are believed to be involved in the contraction of follicles and ovary and β^2-receptor stimulation causes relaxation. The direct effects of neurotransmitters, and their agonists or antagonists on ovulation have produced the better evidence in regard to role of nerves and catecholamines in ovulation (see Goetz *et al.*, 1991 for references). Both α-and β-adreno receptor mechanisms are involved in gonadotrophin-induced ovulation. The increase in cAMP concentrations observed several hours prior to ovulation to have any direct effects on follicular rupture or oocyte expulsion and may instead be involved in setting in motion a cascade of metabolic events that result in ovulation (reviewed by Goetz *et al.*, 1991). Catecholamine-induced increases in cAMP apparently provide a link between their effects and those of other agents as observed by the stimulation of cAMP and progesterone production in follicular tissue with β-adrenoreceptors stimulation (Guraya, 1997a, 2000a). The presence of other neuropeptides such as vasoactive intestinal peptide (VIP) and substance P have also been related to the direct or indirect regulation of steroidogenesis and this possibly occurs via cAMP-mediated mechanism (Guraya, 2000a). Very little or no attempt is made previously to study innervation of human ovary as well as to relate directly the neuro transmitters to ovulation (Guraya, 2000a).

Theca interna constitutes the primary source of follicular relaxin (Bagnell *et al.*, 1987) providing further evidence for a paracrine role for relaxin in the ovulatory process, especially in the contraction of follicle wall. Oxytocin, which is known to be synthesized by the ovary, may also play a role at the local level (reviewed by Thibault and Levasseur, 1988). King *et al.* (1996) have suggested that a higher ovulation rate can be obtained in Merino ewes by repeated low dose I.V. injection of oxytocin during oestrus. A study has also been made of the effect of treatment with a slow-releasing oxytocin preparation at the onset of oestrus on the ovulation rate of this species (King and Coetzer, 1997).

From the presence and synthesis of LH-RH in the ovary, its role in the ovulation is also suggested (Thibault and Levasseur, 1988; Goetz *et al.*, 1991). LH-RH functions primarily by the inositol triphosphate-diacyglycerol proteinkinase-C pathway as it does not mimic the action of gonadotrophins in increasing cAMP concentrations. The action of LH-RH or Gn-RH on follicular rupture at ovulation (Lettrie *et al.*, 1996; Kim *et al.*, 1996) is believed to be related to its ability to increase *in vitro* production of prostaglandin and PA by the granulosa and follicles (Goetz *et al.*, 1991; Tsafriri and Adashi, 1994; Kim *et al.*, 1996) as both are involved in follicular rupture as already discussed. This suggestion is supported by the fact that the inhibitors of prostaglandins inhibit the induction of ovulation by Gn-RH (Goetz *et al.*, 1991; Tsafriri and Adashi, 1994). Bowen *et al.* (1998) have demonstrated the importance of Gn-RH surge for induction of preovulatory LH surge of the ewe; this preovulatory LH surge is not required for induction of ovulation. The precise roles of various protein hormones and neurohormones in follicle rupture and oocyte maturation still remain to be defined more precisely at the molecular level (Goetz *et al.*, 1991; Guraya and Dhanju, 1992; Espey and Lipner, 1994; Guraya, 1997a, 2000a; Fauser *et al.*, 1999; Wolf and Zelinski-Wooten, 2001). Thibault and Levasseur (1988) suggested that some of these peptides may act as amplifier or modulator for gonadotrophin action at the time of the preovulatory surge or may even exert a local negative feed back on their action. Dissen *et al.* (1996) have made a study of the role of trKA nerve growth receptors in mammalian ovulation. The remarkably narrow time frame of trKA gene activation at the completion of follicle growth suggests that NGF action as neuroendocrinotrophic factor in a developmentally restricted manner to the acute cytodifferentiation process that leads to first ovulation in mammals. The experiment carried out by Christensen and Steuffer (1997) demonstrate a novel role for the midcycle surge of gonadotrophin (LH, hCG or FSH) in primates to promote vascular endothelial growth factor (VEGF) by granulosa cells in the periovulatory follicle. FSH-like as well as LH-like gonadotrophins directly stimulate VEGF synthesis by the granulosa cells. The role of VEGF in ovulation is required to be determined.

4.2 CONCLUSIONS : THEIR RELEVANCE TO HUMAN AND RECOMMENDATIONS

In conclusion, it can be stated that the mechanism of ovulation forms a very complex process regulated by close interaction of a variety of chemical factors produced in the follicle wall under the influence of gonadotrophic and steroid hormones. Neural innervation appears to modulate the mechanism of ovulation by producing peptides, which act as amplifiers for gonadotrophic action. The selective use of some of the chemical substances involved in ovulatory process at appropriate times following administration of exogenous gonadotrophins may be helpful in improving the ovulation rate during superovulation in human, primates and domesticated ruminants. It will be interesting to mention here that histochemical studies have demonstrated conspicuous changes in the ovarian stroma and blood vessels with the aging of women (Figs. 7 and 8) (Guraya, 1976b). Some changes are also expected to occur in the nerve supply of ovaries in the aged women, which are required to be investigated with correlative cytochemical, electron microscopic and molecular techniques (Fauser *et al.*, 1999; Tevelde *et al.*, 2000; Bulletti *et al.*, 2001) as the neural innervation is found to modulate the mechanism of ovulation by producing peptides which function as amplifiers for gonadotrophic action in mammals. But in general various nervous components are found to undergo aging changes as supported by molecular changes in mitochondria (Satav and Nair, 1999). Further studies using a variety of techniques as used to understand the mechanism of ovulation at the cellular, molecular and biochemical levels in different mammalian species as discussed above are required to be carried out on the ovaries of aged women as well as under different pathological situations. The results of such studies will be very rewarding to know that to which extent the mechanism of ovulation in human is affected in terms of release of egg from the preovulatory follicle with the aging of women in the normal course of ovulation as well as in response to exogenous gonadotrophins which cause superovulation (Section 4.3). If the release of the egg (s) is delayed it may also cause some aging changes in them, which can affect fertilization *in vivo* and *in vitro* and also embryogenesis (Fauser *et al.*, 1999; Wolf and Zelinski-Wooten, 2001). But simultaneous with expansion of *in vitro* fertilization (IVF) there has been the development of various new approaches such as the assisted reproductive technology (Wolf and Lanzendorf, 1991; Wolf and Zelinski-Wooten, 2001). These will not be described here. Inspite of impressive growth and expansion of these technologies, our knowledge about the physiology and cellular and molecular biology of ovulation in human is still limited due to ethical and practical problems. Thus controlled cellular and molecular investigations that jeopardize the viability of human eggs or embryos are largely precluded, thus indicating the need for the development of various clinical technologies which are based on model mammalian studies; however, such is not always the case because the frontiers for assisted reproductive technologies include the effective treatment of male infertility using micromanipulation and assisted fertilization; maturation of ovarian oocytes and their cryopreservation and the genetic diagnosis of preimplantion stage embryos (Wolf and Zelinski-Wooten, 2001 chapters 5 and 6). Filicori *et al.* (1999) have suggested that low-dose hCG therapy can improve sensitivity to exogenous FSH in patients with secondary amenorrea. But this increased response can be associated with complications such as multiple gestation. Gallowary *et al.* (2000) have suggested that mutations in an oocyte-derived growth factor gene (BMPI5) cause increased ovulation rate and infertility in a dose sensitive manner, suggesting that BMPI5 is essential for female fertility and that natural mutations in a an ovary-derived factor can cause both increased ovulation rate and infertility phenotypes in a dose-sensitive manner.

4.3 OVARIAN HYPERSTIMULATION AND EGG COLLECTION AND QUALITY IN HUMAN

Although limited success is achieved in the recovery and fertilization of eggs from the natural cycles (there is great current interest in returning to natural cycles), most of these studies use controlled ovarian hyperstimulation to obtain the potential number of eggs available to permit a predictable intervention time for their surgical pickup. (Ragni *et al.*, 1999; Weldenstrom *et al.*, 1999; Wolf and Zelinski-Wooten, 2001). Controlled ovarian stimulation during an IVF cycle usually produces large number of oocytes, and consequently it is likely that more embryos will be generated than can be transferred in a given cycle. It is therefore desirable to freeze bank surplus oocytes before insemination to avoid the ethical and legal complications of disposing of or storing embryos.

Two regiments for enhancing gonadotrophic hormone concentrations have been used in recruiting and/or supporting the growth of multiple ovarian follicles; stimulating endogenous hormone release and administering exogenous hormone (Diedrich *et al.*, 1988; Kebly *et al.*, 1999; Shinetugs *et al.*, 1999; Lenton *et al.*, 2000; Felberbaum *et al.*, 2000; Wolf and Zelinski-Wooten, 2001). But limited ovarian stimulation is also repeated. El-sheikh *et al.*, (1999) and Develioglu *et al.*, (1999) have demonstrated the value of basal serum FSH, LH and oestradiol concentrations following pituitary down-regulation in predicting ovarian response to stimulation with highly purified FSH, suggesting that oestradiol and FSH concentration after down regulation are productive of the pattern of ovarian response to stimulation and oocyte yield controlled ovarian hyperstimulation is considered a stalled neutrophil activation (Orvieto *et al.*, 1999) as ovarian hyperstimulation leads to neutrophil activation which correlates with the degree of luteinization. Further studies are required to determine the relationship between the immune system and controlled ovarian hyperstimulation. The results obtained by Shinetuge *et al.* (1999) suggests that the intraovarian colony stimulating factor-I, possibly induced by LH/hCG, plays an important role during ovulation and luteinization. Blasch *et al.* (2000) and De Placido *et al.* (2000) have found that recombinant FSH (res-FSH) is more efficacious than urinary FSH (u-FSH-Hp) when used in the same patient in inducing multiple follicular development in down-regulated cycles as indicated by ovarian performance and oocyte maturity (Schots *et al.*, 2000a, b). In addition, res-FSH yields significantly higher implantation rates than u-FSH-Hp used in-patients undergoing their IVF attempt. Anserind *et al.* (2000) have made a clinical study of a new subcutaneous, purified, urinary FSH preparation for controlled ovarian hyperstimulation in IVF. The results obtained by Cahil and Hull (2000) have suggested an inherent disorder of follicular function, with LH surge impairment probably being a secondary phenomenon. The resulting reduction in the chance of fertilization of the oocyte would contribute substantially to the subfertility associated with endometriosis (see also Cahill and Hull, 2000; Frydman *et al.*, 2000; Garrido *et al.* 2000). It appears that the benefit of IVF is gained through the excessive number of oocytes obtained by stimulation in various studies (Wolf and Zelinski-Wooten, 2001). The diffusion of exogenous gonadotrophin into the FF is considered to be an important predictor of IVF outcome (Nagata *et al.*, (1999). Clomiphene citrate representing the former approach has been used mostly in the clinical management of menstrual regularity and employed extensively during early IVF trials. The drug is known to compete with oestradiol for the latter's receptor and causes an oestrogenic effect on the pituitary. The administration of clomiphene citrate during assisted reproductive technologies is associated with moderate ovarian hyperstimulation and the recovery of three to five eggs. The medication is also employed in conjunction with exogenous gonadotrophins. It will be rewarding to carry out such studies on the comparative aspects of ovarian hyperstimulation in young and aged women.

In the early 1980s, the first IVF programme using pergonal (Serono) a combination of LH and FSH recovered from the urine of postmenopausal women, was initiated in USA (Wolf and Lanzendorf, 1991). Metrodin (Serono) a purified preparation of FSH has also been used more recently in combination with pergonal. These hyperstimulation approaches are used in the follicular phase of the cycle and usually extend for 8 to 10 days. When follicles attain size (15-20 mm mean diameter depending upon the ovarian hyperstimulation protocol and the means of detection) and the circulating urinary concentrations of oestradiol peak or begin to plateau, a bolus administration of hCG is injected to induce preovulatory maturation of the follicle and its enclosed oocyte. A major drawback in these ovarian hyperstimulation has been a high dropout rate related to either an inadequate response or the occurrence of an endogenous LH surge-successful egg collection can be obtained following the occurrence of an endogenous LH surge if frequent monitoring permits a definitive identification of surge initiation. Gull *et al.* (1999) have confirmed *in vivo* anaerobic glycolysis in gonadotrophic, hyperstimulation human ovarian follicles. Chian *et al.* (2000) have investigated whether the rates of oocyte maturation, fertilization and development, as well as pregrancy rate could be improved by hCG priming 36 h before in mature oocyte retrieval in patients with polycystic ovarian syndrome (Sills *et al.*, 2000a,b). There were five clinical pregrancies (37.5%) in the hCG-primed group. Zullo *et al.* (2000) have used minilaparoscopic ovarian drilling under local anesthesia in patients with polycystic syndrome.

GnRH agonists have been used to down regulate pituitary that is to induce hypophysecomy, thereby initiating an endogenous surges; ovarian hyperstimulation and ovulation then become entirely dependent on the injection of exogenous hormone. Down-regulation with GnRH agonists can lead to prolonged exposure to drug, i.e., complete pituitary desensitization before the administration of exogenous gonadotrophins, or a short "burst" protocol in which agonists and exogenous gonadotrophins are injected simultaneously starting on day 3 of the treatment cycle (Wolf and Lanzendrof, 1991; Wolf and Zelinski-Wooten, 2001). GnRH agonist treatment is maintained until hyperstimulation becomes complete and exogenous hCG is injected. This approach has caused not only a significant increase in the percentage of initiated cycles carried out through to oocyte pickup but also to an enhancement in the implantation rate per embryo transfer and in pregnancy rate. A disadvantage of GnRH agonist is the cost related to the enhanced requirements (nearly two fold) for exogenous gonadotrophins. The primary mechanism involved in the preovulatory release of GnRH among induced ovulatory involves the activation of midbrain and brain stem noradrenergic neurons (Bakker and Baum, 2000). GnRH antagonist centrorelix in a single dose administration could represent a first choice IVF treatment with hyperstimulation protocols, and acceptable success rate (Rongieres-Berirand *et al.* 1999b). The use of recombinant gonadotrophins and genetically engineered hCG is believed to be of considerable help for management of low responders (LR) (Pellicer *et al.*, 1998; Penarrubia *et al.*, 1999; Wolf and Zelinski-Wooten, 2001). Similarly, non-gonadotrophin hormones such as cytokines or growth factors, may play a role in the stimulation of ovary (Pellecer *et al.*, 1988; Fauser *et al.*, 1999). A spontaneous cycle and the GnRH antagonist cetrorelix in a single dose administration is believed to represent a first-choice IVF treatment with none of the complications and risks of current controlled ovarian hyperstimulation protocols, and acceptable rate (Rongieres-Bertrand *et al.*, 1999). Intra-venous albumin has been used to prevent severe ovarian hyperstimulation syndrome (Aboulghar *et al.*, 2000). But the results of this review are believed not to be conclusive as they are based on three small trials and thus further studies are required to be carried out. Chen *et al.* (2000b) have studied the role of serum and follicular fluid (FF) pro-inflammatory cytokines and vascular endothelial growth factor (VEGF) in the prediction of ovarian hyperstimulation syndrome (OHSS). This study suggested that FF IL-6 concentrations at the time of oocyte retrieval and serum IL-6 concentrations on the day of embryo transfer might serve as early predictors for this syndrome. Lair *et al.* (1999) have made prediction of severe ovarian hyperstimulation syndrome by free serum vascular endothelial growth factor concentration or the day of hCG administration. Transforming growth factor beta (TGF beta I) in ovarian FF following ovarian stimulation and *in vitro* fertilization correlates to pregnancy (Fried and Wramshy, 1998), showing TGF beta I may be important for successful human pre-embryo development contributes to successful embryo implantation and development and may be necessary for the establishment of pregnancy. Oosterhuis *et al.* (1998) have produced new evidence that IGF-I concentrations in FF is higher in women who respond better to follicular stimulation, i.e., women who grow many follicles, women who need a shorter duration of stimulation and women who need fewer ampeules FSH before oocyte retrieval. Brown *et al.* (2000) and Galloway *et al.* (2000) have suggested that mutations in an oocyte-derived growth factor gene (BMP15) cause increased ovulation rate and infertility in a dose-sensitive manner, establishing that BMP15 is essential for female fertility and that natural mutations in overage-derived factor can cause both increased ovulation rate and infertility pheno types in a dose sensitive manner.

By using the various protocols or agents it will be very rewarding to make comparative studies on ovarian hyperstimulation of young and aged women for determining the number of eggs obtained and their quality for IVF and early embryogenesis (Wolf and Zelinski-Wooten, 2001). Check *et al.* (2000) have reported three pregnancies despite elevated serum FSH and advanced age. The precedents set in these cases can help physician-patient consultation when patients inquire whether there is certain critical FSH concentration above which pregnancy is not possible or an age over which successful pregnancy could not be achieved even of ovulation despite ovarian failure was possible. Check *et al.* (2000b) have suggested that reduced pregnancy rates in older women undergoing ovarian stimulation are more likely to occur because of the oocyte quality rather than uterine senescence. If future studies determine that the use of ovarian stimulation does decrease implantation rates more in older versus younger women; a mechanism other than increased uterine vascular impedance must be sought. Lewit and Kol (2000) have stated that low responder female IVF patient with hypogonadotrophins hypogonadism do not give up.

The collection of mature eggs is important for the success of IVF-ET and thus various picks up techniques have evolved substantially over the last some years (Wolf and Lanzendorf, 1989; Wolf and Zelinski-Wooten, 2001) which will not be described here as these mostly involve physical methods. Semba *et al.* (2000) have reported an autopsy case of ovarian hyperstimulation syndrome with massive pulmonary edema and pleural effusion, recognizing the fact that massive pulmonary edema can occur in a patient with ovarian hyper stimulation syndrome, which is most serious complication of ovulation induction with exogenous gonadotrophins. Paoloni-Giacobino *et al.* (2000) have reported a case of 45, X Turner syndrome with spontaneous ovulation proven by ultrasonography.

5

Fertilization

Following several years of pinoneering work by Patrick Steptoe and Robert Edwards, clinical *in vitro* fertilization (IVF)-embryo transfer (ET) was initiated with the birth of baby Louise Brown in 1978. Human IVF-ET has evolved into standard medical practice for the treatment of many types of infertility (Feichtinger and Kemeter, 1987; Bavister, 1991; Wolf and Lanzendorf, 1991; Englert *et al.*, 1999; Staessen *et al.*, 1999; Kerjean *et al.*, 1999; Witz *et al.*, 1999; Fauser *et al.*, 1999; Yen *et al.*, 1999; Shapiro and Jones, 2000; Aytoz *et al.*, 2000; Tevelde *et al.*, 2000; Meinzo *et al.*, 2000; Bulletti *et al.*, 2001; Wolf and Zelinski-Wooten, 2001). The application originally confined to the hopelessly infertile patient with irreparably damaged or absent fallopian tubes, has advanced to include idiopathic male endometriosis, cervical, and immunologic etiologies. About 15% of egg retrieval cycles result in pregnancy and live birth resulting from the transfer of on overage three embryos. Experience base with IVF-ET in man (Fechtinger and Kemetor, 1987; Bavister, 1991; Wolf and Lanzendorf, 1991; Englert *et al.*, 1999; Staessen *et al.*, 1999; Kerjean *et al.*, 1999; Fauser *et al.*, 1999; Yen *et al.*, 1999; Witz *et al.*, 1999; Shapiro and Zones, 2000; Autoz *et al.*, 2000: Tevelde *et al.*, 2000; Meinzo *et al.*, 2000; Bulletti *et al.*, 2001; Wolf and Zelinski-Wooten, 2001) exceeds than in all other mammalian species with the possible exception of the mouse (Gosster, 1992). Simultaneously with the expansion of IVF there has been the development of new approaches such as the assisted reproductive technologies which include gamete intrafallopian transfer (GIFT); zygotic intrafallopian transfer (ZIFT); pronuclear stage transfer (PROST); tubal embryo transfer (TET); superovulation uterine replacement capacitation enhanced sperm (SOURCE); the direct aspiration of epididymal sperm for IVF, the freezing of embryos, and the use of micromanipulaion to assist fertilization as well as transfer (Feichtinger and Kemeter, 1987; Van Blerkom, 1989; Wolf and Lazendorf, 1991; Dunbar and O, Rand, 1991; Rajalakshmi and Griffin, 1999; Wolf and Zelinski-Wooten, 2001) are of current development. Inspite of the impressive growth and expansion of the assisted reproductive technologies in relation IVF-ET, our knowledge about the and cell and molecular biology associated with human fertilization is still limited due to ethical and practical problems (Feichtinger and Kemeter, 1987; Bavister et al., 1990; Fauser *et al.*, 1999; Yen *et al.*, 1999; Tevelde *et al.*, 2000; Meinzo *et al.*, 2000; Bulletti *et al.*, 2001; Wolf and Zelinski-Wooten, 2001). Thus, controlled and correlated cellular and molecular studies that jeopardize the viability of human eggs and embryos are precluded suggesting that the development of various clinical technologies is based on model mammalian studies; however, such is not always the case.

The study carried out by Neuber and Powers (2000) has suggested that the mouse is not a clinically relevant model for human fertilization failures.

Frontiers of the assisted reproductive technologies in 1989 include the effective treatment of male infertility using micromanupulation and assisted fertilization; maturation of ovarian oocytes and their cryopreservation; and the genetic diagnosis of preimplantation-stage embryos (Feichtinger and Kemeter, 1987; Van Blerkom, 1989; Menoza and Ben-Khalifa, 1995; Wolf and Zelinski-Wooten, 2001). IVF provides evidences for the heterogeneity of human embryos; (Gregory, 1998); the data from U.K. indicating that approximately 90% of embryos selected for transfer fail to implant. These is now increasing awareness that the heterogeneity of follicles may have a significant impact on oocyte competence and embryo viability and that factors which contribute to the heterogeneity of follicles may provide markers of implantation in assisted conception (Fauser *et al.*, 1999; Yen *et al.*, 1999; Tevelde *et al.*, 2000; Bulletti *et al.*, 2001). The markers of implantation described include the expression of beta-hydroxy steroid dehydrogenase by granulosa cells in vitro; adhesion and proliferation of cumulus cells *in vitro*, steroidogenic activity of cumulus *in vitro* and perifollicular vascularity and vascular endothelial growth factors bound to granulosa and cumulus cells. These factors may provide clinically useful markers of implantation potential.

5.1 FUSION OF SPERM AND EGG

The fertilization involves fusion of sperm and egg and it is important to establish the diploid complement required for the development of embryo in human and other mammalian species (Zones 1984; Cooper, 1986; Dietl, 1989; Feichtinger and Kemeter, 1987; Dunbar and O, Rand, 1991; Van Blerkom, 1989; Fauser *et al.*, 1999; Yen *et al.*, 1999; Tevelde *et al.*, 2000; Meinzo *et al.*, 2000; Bulletti *et al.*, 2001; Wolf and Zelinski-Wooten, 2001). Fertilization in human and other mammalian species follows a series of defined events that occur in succession before fusion of gametes occurs (Figs. 18, 19 (A-H)). The chronological order in which these events occur is well defined and is as follows: (1) A loose nonspecific association between sperm and zona pellucida (ZP) of the egg and is called as attachment. (2) Attachment is followed by a more tenacious association of sperm with ZP of the egg called as binding, which is mediated, by the egg. (3) The bound sperm then undergoes a distinct morphological change called acrosome reaction that involves fusion and vesiculation of sperm outer acrosomal membrane resulting in the release of acrosomal contents and exposure of inner acrosomal membrane (Gossler, 1992; Fauser *et al.*, 1999; Guraya, 2000b). (4) Sperm acrosome reaction helps the sperm in penetration of the ZP possibly involving acrosomal hydrolases in the process. (5) After penetration, sperm reaches the perivitelline space. "Fusion" occurs between the sperm plasma membrane at the postacrosomal region and egg plasma membrane. Sperm-egg fusion appears to be less species specific-than sperm-zona interaction. Fusion precludes fusion of additional sperm and thus inhibits polyspermy, possibly by depolarizations of egg plasma membrane that induces cortical reaction, which involves fusion of cortical granule membrane with the egg plasma membrane. The contents of cortical granules are released into the perivitelline space, which affects the ZP (zona reaction) (Figs. 18, 19 (C-D)). This hardening of ZP (zona reaction) inhibits polyspermy (Guraya, 1985). (6) Fusion with single sperm activates the egg and embryonic development starts immediately (Glosser, 1992). A comparative study is required to be made of all these events at the cellular and molecular levels for the fertilization of eggs of young and aged women (Fauser *et al.*, 1999; Yen *et al.*,

1999; Tevelde *et al.*, 2000; Meinzo and Barak, 2000; Bulletti *et al.*, 2001), which may help us to find out if there some disorders in the processes of their fertilization and subsequent early embryogenesis in aged women as the primordial oocytes undergo aging changes (Guraya, 1999a). Such studies can be carried out *in vitro* systems (Feichtinger and Kemeter, 1987; Wolf and Zelinski-Wooten, 2001). Magerkurth *et al.* (1999) scanning electron microscopy analysis of the human zona pellucida have investigated the influence of maturity and fertilization on morphology and sperm binding pattern. In both the mature and immature oocytes, there were some oocytes with either no or numerous bound spermatozoa on the ZP. Oocytes overloaded with spermatozoa could only be found in the mature group. Unfertilized oocytes had fewer bound spermatozoa on average than polyploid zygotes.

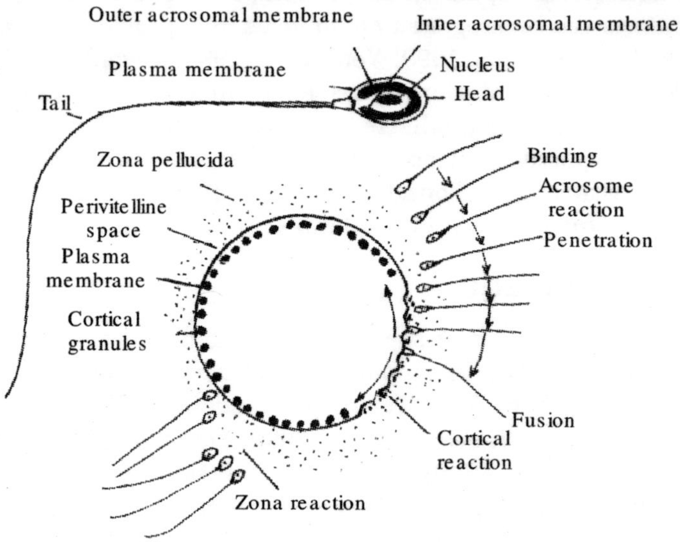

Fig. 18

5.2 EPIDIDYMAL AND EJACULATED SPERM

As a rule for the epididymis in sperm maturation, which is well established in rodents, is inferred in man from investigations associating the acquisition of motility, fertility potential as measured with the sperm penetration assay and *in vivo* fertility in the epididymal exposure. And rogen-induced proteins are reported that coat sperm during epididymal transit, and fertility potential is enhanced in sperm obtained more distally from successive epididymal segments, *i.e.*, from the caput to the cauda (Bavister *et al.*, 1990; Rajalakshmi and Griffin, 1999; Liu *et al.*, 2000). A significant role for the epididymis in sperm maturation and storage is still being debated as in the epididymo-vasectomy patients the sperm exposure to only the caput epididymis is found to be sufficient for maturation; some workers have successfully aspirated fertile sperm from the proximal caput epididymis in patients with corgenital absence of the vas deferens (Feichtinger and Kemeter, 1987; Wolf and Lanzendorf, 1991; Rajalakshmi and Griffin, 1991).

A normal human ejaculate shows relatively wide range in sperm motility and morphology (Feichtinger and Kemeter, 1987; Van Blerkom, 1989; Wolf and Lanzendorf, 19991; Dunbar and O, Rand, 1991; Rajalakshmi and Griffin, 1999). In contrast with other mammals, the normal human ejaculate is "dirty". Besides its cellular components, ejaculates show complex secretions from the accessory glands-the Cowper's gland, prostate, and seminal vesicles which, in turn, show a variety of energy substrates, hormones, nonenzymatic and enzymatic proteins and various ions (Cooper, 1986; Feichtinger and Kemeter, 1987; Rajalakshmi and Griffin, 1999; Fauser *et al.*, 1999). Within the first few minutes following ejaculation, semen liquefaction in human takes place due to enzymatic activities originating from prostatic secretions. Therefore, a prolonged sperm exposure to seminal plasma *in vitro* leads to detrimental effects on motility and viability. *In vivo*, sperm are separated from the seminal plasma within minutes of coitus; on colonization of cervix and its mucus.

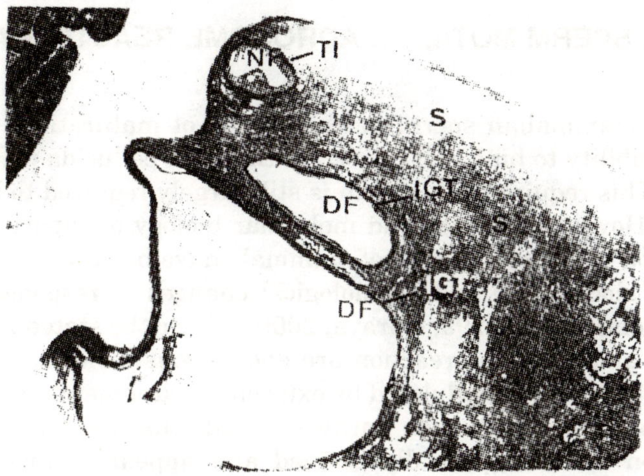

Fig. 19

5.3 CONVENTIONAL SEMEN ANALYSIS

Conventional semen evaluation constitutes a mainstay of clinical work up for male infertility. The analysis provides a description of the ejaculate volume, viscosity, and colour as well as a description of the cellular fraction. The World Health Organization has produced normal values for sperm concentration, motility, progression and morphology (Wolf and Lazendorf, 1991). A role of electron microscopy in the analysis of semen quality is also advocated (Feichtinger and Kemeter, 1987; Zamboni, 1987; Demartino *et al.*, 1989). Advances in technology support the quantitation of motility by computer-assisted video image analysis. Unfortunately, observations made on conventional semen analysis show a poor correlation with fertility except in extreme cases such as azoospermia and/or as the nospermia. This absence of correlation may suggest that the endpoint is crude when one takes into consideration the number of physiological and biochemical observations which occur during sperm transit in the female reproductive tract coupled with the needs that only a few (one) capacitated sperm present at the right time and the right place adequately result in fertility. In regard to IVF outcome it becomes immediately clear that the total number of morphologically normal, motile sperm needed to obtain maximum fertilization levels is greatly decreased over that required *in vivo* (Feichtinger and Kemeter, 1987 Fauser *et al.*, 1999; Wolf and Zelinski-Wooten, 2001). The methods used for semen analysis, of course, depend on the desired use. For cervical in semination whole semen is used. For cryopreservation, liquefied semen can be directly diluted with cryoprotectant or with cryoprotectant plus extender (s) in preparation for freezing (Feichtinger and Kemeter, 1987; Wolf and Lanzendorf, 1991; Wolf and Zelinski-Wooten, 2001). For the *in vitro* separation of sperm from seminal plasma, a number of methods are being used ranging from a simple swim-up from whole semen or a washed sperm pellet to the use of buoyant density gradient centrifugation approaches involving commercial products such as Percoll or Sperm Select. The success in the amplification and analysis of DNA sequences in a single sperm is of great help in the genetic analysis of the male contribution to fertilization and development (Li *et al.*, 1988; Van Blerkom, 1989; Fauser *et al.*, 1999; Wolf and Zelinski-Wooten, 2001).

5.4 CAPACITATION, SPERM MOTILITY, ACROSOME REACTION AND SPERM-ZONA BINDING

Freshly obtained mammalian sperm need a period of maturation, called capacitation, before developing the ability to bind and penetrate the zona pellucida (ZP) and fuse with the egg (Guraya, 2000b). This requirement in man is still largely required to be defined precisely (Fauser *et al.*, 1999). However, cellular and molecular biology of capacitation and acrosome reaction in spermatozoa in various species of mammals have been investigated by comparing and contrasting their biochemical and physiological changes in response to various factors *in vivo* and *in vitro* (Fauser *et al.*, 1999; Guraya, 2000b). It can be stated now that phenomena of sperm capacitation and acrosome reaction are endogenous molecular events occurring at the membrane level, which can be modulated by external environmental factors. The molecular mechanisms and the signal transduction pathways mediating the processes of capacitation and acrosome reaction are still partially defined and appear to involve modification of intracellular Ca^{2+} and other ions, lipid transfer and phospholipid remodeling in the sperm plasma membrane as well as changes in protein phosphorylation as discussed in detail by Guraya (2000b). Evidences for the involvement of cAMP-dependent kinase pathway in the acrosome reaction are also discussed. The mediation of one or more external signals by the sperm plasma membrane appears to activate this pathway after or simultaneous with the influx of Ca^{2+}. Concurrent with or following of entry of Ca^{2+}, adenylate cyclase is activated resulting in the increased levels of cAMP-activation of cAMP-dependent kinase and protein phosphorylation. The identity of such proteins and their role in acrosome reaction are required to be determined (Fauser *et al.*, 1999; Guraya, 2000b). The roles of biological effectors of the acrosome reaction such as ZP_3 and the follicular fluid are still to be defined at the molecular level. In future, various extrinsic and intrinsic molecular probes will certainly be very useful to solve the controversies involved in elucidation of the molecular mechanism of sperm capacitation and acrosome reaction.

The assessment of capacitation using human IVF is limited by practical and ethical considerations (Feichtinger and Kemeter, 1987; Van Blerkom, 1989; Wolf and Zelinski-Wooten, 2001). Since no overt morphological correlation still can be evolved to identify capacitation, its measurement generally involves monitoring acrosomal loss (Wolf and Lanzendorf, 1991; Fauser *et al.*, 1999; Guraya, 2000b) or two rather cumbersome sperm-egg interaction systems: penetration of zona enclosed nonviable human oocytes or of zona-free hamster eggs. The disadvantage of these two bioassays is that the performance of only a limited number of sperm can be monitored, needing substantial extrapolation to determine the capacitation status of the entire population. The most important component in initiating capacitation of seminal sperm appears to be removal of the seminal plasma as this process possibly leads to changes of the membrane structure and function. The persistent presence of even dilute seminal plasma inhibits the attachment and penetration of zona-free hamster and salt-stored human eggs (Feichtinger and Kemeter, 1987; Wolf and Lanzendorf, 1991; Fauser *et al.*, 1999). A wide variety of media are found to support the capacitation of human sperm as determined by clinical IVF-use (Wolf and Zelinski-Wooten, 2001). These include both simple and complex bicarbonate-buffered solutions, generally augmented with an albumin source (Guraya, 2000b). Albumin-mediated sperm capacitation involves sterol efflux (Guraya, 2000b); however, it has become amply clear that the capacitation does not depend on albumin as it can be conducted in protein-free or in gelatin (Wolf and Lanzendorf, 1991).

For the past some years, with the availability of automated, quantitative methods for motility measurement (Guraya, 2000b), the concept that hyperactivation forms a component of capacitation has been extended to man. Hyperactivation occurs when seminal or washed sperm, which typically swim with a linear trajectory, change trajectory to a complex nonlinear form (Feichtinger and Kemeter, 1987; Wolf and Zelinski-Wooten, 2001). Hyperactivation appears to show the first overt sign of capacitation, followed by the ability of the capacitated sperm to bind to the zona pellucida and undergo an acrosome reaction (Guraya, 2000b; Wolf and Zelinski-Wooten, 2001). In the beginning, human sperm in capacitating medium were not believed to show hyperactivation although alterations in the type of motilities shown in synthetic media were recognized (Feichtinger and Kemeter, 1987; Wolf and Lanzendorf, 1991; Fauser *et al.*, 1999; Wolf and Zelinski-Wooten, 2001). Subsequently, it was shown that approximately 20% of washed sperm showed hyperactivation. The ovulation and use of a computer-assisted digital image analysis system for the rapid measurement of a range of motility parameters in individual sperm led to the development of an automated method for quantitation of hyperactivation (Robertoon *et al.*, 1988). This transition in trajectory takes place in some sperm immediately on removal from the seminal plasma and possibly in all cases before spontaneous acrosomal loss. The requirements for optionally determining hyperactivation in capacitating populations have been determined, and it seems that the extent of hyperactivation is a donor feature (Wolf and Lanzendorf, 1991; Guraya, 2000b).

Capacitated acrosome intact sperm in many mammalian species are observed to be capable of binding to the zona pellucida (Fauser *et al.*, 1999). Binding forms the essential step for zona penetration as it initiates events that lead to acrosome reaction. A specific zona glycoprotein functions a receptor and as a potent inducer of acrosome reaction (Guraya, 2000b). The human zona pellucida also contains three acidic glycoproteins comparable to those in the mouse ZP_1, ZP_2 and ZP_3 (Glossler, 1992; Fauser *et al.*, 1999; Guraya, 2000a, b; Chapter 2). Moreover, there occurs a fertilization-dependent modification of ZP_1. The kinetics of capacitated sperm binding to the zona have been investigated using nonfertilizable ovarian oocytes (Wolf and Lanzendorf, 1991). Sperm can acquire the ability to bind and penetrate the zona rather quickly (within several hours) but surprisingly this ability is lost after 20 hr of “capacitation”. Both acrosome intact and acrosome reacted sperm have been observed on the zona surface by transmission electron microscopic examination (Van Blerkom, 1989) zonae discarded from IVF attempts have also been studied to measure fertility potential (Wolf and Lanzendorf, 1991; Fauser *et al.*, 1999; Wolf and Zelinski-Wooten, 2001; Section 5.4).

A physiological acrosome reaction is observed to occur only in a capacitated sperm (Guraya, 2000b); thus, the ability of free-swimming sperm to undergo acrosome reaction is equated with capacitation or fertility potential (Guraya, 2000b). The development of indirect immunofluorescence methods at the light microscope level is found to be helpful to quantitate acrosomal status in large populations of sperm rapidly and accurately. Unlike the rodent, human sperm incubated under capacitating conditions do not show spontaneous loss of acrosome *in vitro* (Fauser *et al.*, 1999; Wolf and Zelinski-Wooten, 2001); the reacting population may be restricted to only 10% of the motile cells. Moreover, sperm do not show degenerative acrosomal loss as an important component of senescence or decay. Acrosomal loss can be achieved by biological agents such as follicular fluid, cumulus cells, or zona pellucida or by physiological agents such as calcium ionopheres, lysophosphatidylcholine, and electropermeabilization (Guraya, 2000b). The study of capacitated cells to respond to ionophore induction of acrosomal loss represents convenient assay for capacitation (Guraya, 2000b). Washed motile sperm incubated under capacitating conditions can definitely be distinguished (based on their state of capacitation) into inducible and non-inducible populations. By employing ionophore-induced acrosomal loss as a measure, the kinetics of capacitation can be determined. The responsive population increases in size to a maximum value of 35% of the motile cells within 6-8 hr of incubation. These conclusions are derived mainly from studies on free-swimming sperm populations. In the fertilizing or zona-penetrating sperm, the acrosome reaction appears to occur after zona binding as induced by ZP_3, in analogy to the mouse (Gossler, 1992; Fauser *et al.*, 1999; Guraya, 2000b). Evidence to support this sequence has been produced in man based on an assessment of acrosomal status of sperm associated with the zona pellucida (Cross *et al.*, 1988; Van Blerkom, 1989; Fauser *et al.*, 1999; Wolf and Zelinski-Wooten, 2001). When sperm and eggs are coincubated for very short periods, almost all attached sperm showed acrosome intact. In contrast, both, acrosome-intact and acrosome-reacted, sperm were seen bound to the zona when a 60-min coincubation period was used. In this investigation, the acrosome reaction could also be produced by acid-disaggregated zonae. Functional assays with auti-ZP_3 synthetic peptide antibodies have indicated that the antiserum does not inhibit sperm-ZP binding whereas one of the antiserum against synthetic ZP_2 peptides significantly inhibits binding of spermatozoa to the ZP.

At the molecular level, both capacitation and the acrosome reaction depend on the presence of exogenous calcium, showing that enhanced intracellular concentrations of this cation trigger the membrane fusion events associated with the acrosome reaction as discussed in detail by Guraya (2000b). Clearly the ability to induce acrosomal loss with calcium ionophores involves calcium as an intracellular messenger. Direct evidence is now produced documenting an increase in intracellular calcium as a concomitant of induced acrosomal loss (Guraya, 2000b).

5.5 FERTILITY POTENTIAL ASSESSMENT

Although the zona pellucida forms the major species-specific barrier to fertilization, in many cases the zona free egg also maintains specificity (Feichtinger and Kemeter, 1987; Fauser *et al.*, 1999). The molecular basis of this zona free egg specificity is required to be determined. But it may be functioning at the level of oolemma which undergoes molecular changes during oocyte growth (Guraya, 2000b; Chapter 2). But hamster forms, a major exception to this rule as capacitated sperm of many mammals can fuse with the zona-free hamster egg (reviewed by Wolf and Lanzendorf, 1991, Fauser *et al.*, 1999). This result led to the development of a sperm penetration assay (SPA) using hamster eggs in the diagnostic evolution of fertility potential in man. Clinical application of the SPA can be predicted on the basis of differences in test outcome found with sperm from males of proven fertility as compared to infertility patients in which the female partner did not show any evidence of infertility. During the last several years correlations have also been made and established between test scores and *in vitro* fertilization of wives eggs (Feichtinger and Kemeter, 1987; Wolf and Zelinski-Wooten, 2001). Besides extensive application, the useful application of the SPA in directing clinical treatment is being still questioned (Mao and Grimes, 1988). Although the test may assess sperm capacitation and the acrosome reaction, it is obviously neither the optimum nor the ideal bioassay because it assesses the relevance of sperm penetration of heterologous eggs to homologous fertilization, and the sperm fusion with the zona-free egg is not always indicative of the ability to penetrate the zona intact egg. Over and above these theoretical arguments, the methodology for carrying out the SPA has varied greatly. As originally reported and applied sperm were capacitated in albumin containing physiological saline solutions for variable time periods at 37°C. This methodology resulted in penetration indices (mean number of penetrated sperm inseminated egg) of less than 1. In order to enhance the sensitivity of the assay, creative methods for sperm capacitation have been introduced, including exposure to the calcium ionophore A 2318, to strontium, or to TES-Tris egg yolk buffer at 4°C for prolonged time periods (reviewed by Wolf and Lanzendorf, 1991; Guraya, 2000b). The latter approach is correlated to high penetration indices (greater than 6 and as high as 50) and has enhanced assay reliability. Since the test depends on the presence of acrosome-reacted sperm, these creative capacitation methods possibly lead to enhanced levels of visible acrosome-reacted sperm. This expectation is realized; however, relatively small alterations in acrosomal-status stand opposed to large changes in mean penetration indices (Wolf and Lanzendorf, 1991; Fauser *et al.*, 1999). Capacitation kinetics, based on SPA results, is donor specific. In some cases functional competence occurs within 3-6 hr of sperm removal from seminal plasma, whereas in others it takes more than 10 hr.

Sperm kinding to human zonae pellucida is used as a means of testing fertilizing potential of individual with possible of male factor infertility (Feichtinger and Kemeter, 1987; Alexander *et al.*, 1989; Wolf and Lanzendorf, 1991; Fauser *et al.*, 1999; Wolf and Zelinski-Wooten, 2001). Variant on this theme, hemizona assay (HZA) utilizes discarded or surplus nonliving, nonfertilizable oocytes obtained during IVF. With micromanipulative methods zona pellucidae are bisected into equal halves, one half to be inseminated with doner sperm (Feichtinger and Kemeter, 1987). Sperm from men with IVF failure showed a significantly lower binding property to hemizonae as compared to the sperm of fertile man. The poor binding of sperm is believed to result from morphological abnormalities or from interfering activity of substances in seminal plasma such as microorganisms or antibodies. Using salt-stored zona pellucida from human oocytes that failed to fertilize *in vitro* it is also observed that individuals showing poor zona-binding ability frequently showed poor or failed IVF (Fauser *et al.*, 1999; Wolf and Zelinski-Wooten, 2001). In this investigation fluorochrome makers were employed to differentiate test sperm from those already bound as a result of the IVF attempt. The involvement of progesterone in the physiological process leading to fertilization of the oocyte has been suggested in several studies (Baldi *et al.*, 1999). In particular the demonstration that sperm responsiveness to progesterone is impaired in subfertile patients and that is strictly correlated to the ability of fertilizing the oocyte represents a further indication of the participation of the steroid in this process. In addition, the determination of sperm responsiveness may be predictive of fertilizing ability with a positive predictive value of 90% and can be clinically useful for the preliminary assessment of the male partner to select the appropriate assisted reproductive technique.

5.6 INSEMINATION (FERTILIZATION)

In vivo the egg after ovulation is transported into the ampulla, the anteriormost part of the oviduct (Fig. 2D). This is achieved by the movements of the cilia of the epithelial cells lining the opening of the oviduct facing the ovary (infundibulum). The egg is surrounded by its zona pellucida (ZP) and accompanying cumulus cells which are embedded in a matrix of proteins and hyaluronic acid. To achieve insemination or fertilization, the sperm must first penetrate the viscous matrix of hyaluronic acid surrounding the cumulus cells and egg. This might be aided by hyaluronidase, a hyaluronic acid hydrolyzing enzyme which is present in the sperm fluid. Sperm associates with the surface of the ZP and then binds more tightly in a species-specific manner (Figs. 18, 19A, 19B). This binding is brought about by interactions between egg binding proteins of the sperm head and the zona component ZP_3, which is the sperm receptor. Binding to ZP_3 elicits the acrosome reaction (as already discussed in Section 5.3): the outer membrane of the acrosomal vesicle fuses with the overlying sperm plasma membrane at multiple sites. The fusion results in the formation of numerous vesicles, which are ultimately shed, exposing the inner acrosomal membrane to the outside (Fig. 21) (Guraya, 2000b). Proteolytic and glycolytic enzymes are released permitting the sperm to penetrate through limited proteolysis. One of these enzymes acrosin has a structure similar to trypsin, is activated after the acrosome reaction (Guraya, 2000b). Acrosin appears to remain closely associated with the sperm head. After, penetrating the ZP, the posterior part of the sperm head fuses with the egg membrane and triggers a cascade of events which prevent polyspermy and ultimately result in the formation of diploid zygote.

By using a wide variety of culture media for sperm preparation and capacitation as already discussed (Section 5.3) oocyte maturation, insemination and embryo development in human IVF are being successfully carried out (Feichtinger and Kemeter, 1987; Wolf and Lanzendorf, 1991; Kuleshova *et al.*, 1999; Fauser *et al.*, 1999; Wolf and Zelinski-Wooten, 2001). All these media range from simple salt solutions to complex media having amino acids and vitamins. Protein supplementation is common and its sources are the maternal or human foetal cord serum and human or bovine serum albumin. Human IVF has also been carried out successfully in defined media (Feichtinger and Kemeter, 1987; Wolf and Lanzendorf, 1991; Fauser *et al.*, 1999; Wolf and Zelinski-Wooten, 2001). The culture media used also have energy sources such as pyruvate and lactate for sperm, oocytes and pronuclear embryos and pyruvate and glucose for cleaving embryos. Based on glucose turnover experiments, Walls *et al.* (1987) have observed that before day 3 (eight cells) of development, human embryos do not effectively use glucose as an energy source because of a blockage in glycolysis. These observations have provided a basic for the inclusion of lactate and/or pyruvate in media used for culturing human embryos during the first cleavage divisions (Feichtinger and Kemeter, 1987; Fauser *et al.*, 1999; Wolf and Zelinski-Wooten, 2001). Typical culture methods also require incubation of gametes and embryos in a 5% CO_2 in air atmosphere at 37°C with a pH ranging from 7.2 to 7.5 and osmolarity of 280 to 305 m Osm/kg. Yao *et al.* (1999) by following the criteria of motility, acrosome reaction, zona binding, and oocyte fusion, have demonstrated that oviductal cells maintained the fertilizing capacity of spermatozoa, whereas human follicular fluid facilitated the fertilization process of oviductal spermatozoa. But the molecular mechanisms involved are required to be determined (Fauser *et al.*, 1999). Espana *et al.* (1999) have observed that protein C present in an active form in follicular fluid inhibits both binding and penetration of zona-free hamster oocytes by human sperm, suggesting that protein C inhibitor may be involved in human reproduction at several steps, including the fertilization process. Administered subcutaneously, the third generation gonadotrophin preparation "Fertinex" is effective *in vitro* fertilization treatment in young women (Crain *et al.*, 1998). According to Newton *et al.* (1999) oocytes that had failed to fertilize had a slightly lower hydraulic conductivity and dimethyl sulphoxide permeability and, that after exposure to 1-5 mol% dimethyl sulphoxide 1-1, these cells appeared to become permeable to normally impermeable solutes. These permeability properties have been used to design a protocol for the addition and removal of dimethyl sulphoxide to control the magnitude of volumetric changes of human mature oocytes.

Fertilization of human oocytes has been successful with sperm concentrations ranging from 1.0×10^4 to 50×10^5 motile sperm per milliliter, with 2.5 to 5.0×10^4/ml as an optimal range for normal fertilization (see Feichtinger and Kemeter 1987; Wolf and Lanzendorf, 1991; Fauser *et al.*, 1999; Wolf and Zelinski-Wooten, 2001). When fertilization is carried out with sperm from a "subfertile male" as observed by abnormal semen parameters, the sperm concentration is often increased to 50×10^4 motile sperm/ml.

The occurrences of polyspermic fertilization have been correlated to sperm concentrations, oocyte maturational status at insemination, the age of the inseminated egg and stimulation protocol used (Feichtinger and Kemeter, 1987; Wolf and Lanzendorf, 1991; Wolf and Zelinski-Wooten, 2001). Polyspermy rates typically vary from 2% to 6%. The differences in the incidence of polyspermic fertilization in the eggs of young and aged women are still required to be determined as the primordial oocytes undergo aging changes (Guraya, 1999a; Chapter 1). The results of such studies will be very rewarding to determine the effects of aging of primordial oocytes on the subsequent problems of fertilization and early development as the quality of eggs must be affected as a result of aging process of primordial oocytes with advancing age of women as discussed by Guraya (1999a) (Chapter 1). However, Kuleshova *et al.* (1999) have reported the birth of a healthy baby girl at 37 weeks gestation to a 47 years old recipient, after vitrification of mature oocytes in IVF.

As reported earlier, the human semen generally contains many morphologically abnormal sperm such as those with large or small heads, double heads, tail defects, midpiece defects, cytoplasmic droplets and immature forms (Guraya, 1987; Feichtinger and Kemeter, 1987; Wolf and Lanzendorf, 1991). The percentage of sperm with "normal" morphology has been correlated with fertilization rate. By employing strict criteria for evaluation, it is observed that fertilization rates are significantly higher for patients whose sperm samples show greater than 14% normal morphology. A normal acrosomal morphology is also shown to be important for successful human IVF. The significance of the acrosome to successful sperm-egg interaction is evaluated from the failure of acrosomeless sperm round-headed sperm syndrome to penetrate hamster eggs (Wolf and Lanzendorf, 1991). Hazzouri *et al.* (2000) after studying genome organization in the human sperm nucleus by FISH and confocal microscopy; have shown a particular distribution of chromosome territories that could he related to mechanisms implicated in its specific functions. The analysis of more chromosomes and chromosomal structure including the Y chromosome will be of help to understand the structure of the human sperm chromatin in the normal and pathological situations and their fundamental and clinical implications.

Fertilization of human oocytes including cryopreserved and thawed oocytes have been carried out by intracytoplasmic normal injection of spermatozoa (Ludwig *et al.*, 1999; Huang *et al.*, 1999; Kupker *et al.*, 1999; Wurfel *et al.*, 1999; Tesarik and Mendozoa, 1999; Goud *et al.*, 1999; Devos *et al.*, 1999; Porcu *et al.*, 1999; Ding *et al.*, 1999; Yanagida *et al.*, 1999, 2000; Weissman *et al.*, 1999; Rawe *et al.*, 2000; Sills *et al.*, 2000; Yamano *et al.*, 2000; Chen *et al.*, 2000; Tsai *et al.*, 2000a; Gvakharia *et al.*, 2000; Kovacic *et al.*, 2000; Wolf and Zelinski-Wooten, 2001) and damaged spermato zoa (Ahmadi and Ng, 1999). Evaluation has also been made of the fertilization potential of freshly isolated, *in vitro* cultured and cryopreserved human spermatids by injection into hamster oocytes (Aslam and Fishel, 1999). Oocyte activation has also been induced by spermatids (Yanagida *et al.*, 2000). Although great progress has been made in both the study and treatment of infertility, a considerable number of patients still fail to conceive because of spermatogenic failure and/or oocyte aging and legal; ethical and even social problems (Tsai *et al.*, 2000b; Wolf and Zelinski-Wooten, 2001). In conjunction with intracytoplasmic sperm injection, fertilization of human oocytes with immature sperm precursors, e.g., spermatids and even secondary spermatocytes, has resulted in healthy babies as already described. Pregnancies have also resulted from the use of spermatids derived from *in vitro* spermatogenesis (Tsai *et al.*, 2000b). The use of human male stem cells might provide an attractive source for the treatment of males with arrested spermatogenesis, as well as male cancer patients (Wolf and Zelinski-Wooten, 2001). It is also suggested that transplantation of somatic cell nuclei and their haploidization within oocytes may prove to be a practical way of eradicating age-related aneuploidy and thus forms an innovative source of healthy oocytes. But the safety of these procedures is required to be proven before their application in the human. Tsai *et al.* (2000b) have also discussed the implications of cytoplasmic quality and genetic imprinting in the context of these manipulations (see also Wolf and Zelinski-Wooten, 2001).

Approximately 12 to 18 hr after fertilization, the fertilized eggs will show two readily visible pronuclei (Fig. 19F). Two polar bodies will also be seen, which may be obscured by cumulus cells or fragmented, and, therefore, should not be used as a major criteria for normal fertilization. Failure of eggs to fertilize *in vitro* may be due to immaturity (cytoplasmic and/or nuclear), postmaturity at the time of insemination or a lack of sperm fertilizing ability (Feichtinger and Kemeter, 1987; Wolf and Lanzendorf, 1991; Fauser *et al.*, 1999 ; Wolf and Zelinski-Wooten, 2001). It will be rewarding to make comparisons of failure of eggs to fertilize, which are obtained from young and aged women as with aging of women, the primordial oocytes undergo cytological, cytochemical and molecular changes (Guraya, 1999; Chapter 1). These changes may affect the capability of eggs to show normal fertilization in older groups of women.

The successful reinsemination of unfertilized oocytes has been described (Wolf and Lanzendorf, 1991). But the efficiency of the method remains to be determined, as it is not certain if such embryos lead to pregnancy following transfer. Visual determination of oocyte nuclear maturation to find the appropriate time for insemination may decrease the requirement for insemination. In an attempt to successfully reinseminate eggs or to treat male factor infertility, Malter and Cohen (1989) have introduced a micromanipulation technique called partial zona dissection. With this technique a microneedle is passed through the zona pellucida to develop an opening through which sperm can enter to form contact with the oocyte plasma membrane. Enhanced fertilization rates after carrying out reinsemination and live/born offspring have formed after partial zona dissection. De Vos and Steirtegham (2000) have discussed new achievements in assisted reproduction with special reference to zona hardening, zona drilling and assisted hatching (Wolf and Zelinski-Wooten, 2001). But five randomized controlled investigations have suggested that assisted hatching is of no benefit to the overall patient population-might be of value in increasing embryo implantation rates only in selected cases. No further evidence exists for an age-related benefit from assisted hatching in patients with advanced maternal age. Hamatani *et al.* (2000) have produced data which suggest that human SP-10 expressed on the equatorial region of acrosome-related sperm, indeed mediates sperm-oolemma binding in a beta (1) integrin-independent manner, but not sperm-zona binding.

As the transfer of multiple embryos enhances the risk of multiple pregnancy, many IVF programmes cryopreserve " surplus" embryos for transfer in subsequent unstimulated menstrual cycles (Feichtinger and Kemeter, 1987; Wolf and Zelinski-Wooten, 2001). These different cryoprotectants used are glycerol, dimethyl sulfoxide and 1, 2 propanediol (Wolf and Lanzendorf, 1991). Besides these freezing media, the freezing vessel used, the speed of the freezing and thawing procedure, and the cell stage at which the embryo is frozen vary between IVF programmes (Kuzan and Quinn, 1998, for review). Cryosurvival is typically around 70%, with approximately 20% of embryo transfers producing a pregnancy.

5.7 CORTICAL ACTIVATION AND THE BLOCK TO POLYSPERMY

Sperm penetration of the mammalian oocyte starts the cortical reaction, an exocytotic fusion event between the oocyte plasma membrane and the cortical granule membrane. Hydrolytic enzymes released from the cortical granules diffuse into the zona pellucida (ZP) and modify its structure. This process is called as the zona reaction which functions in the prevention of polyspermy as already stated (Chapter 2) as it stops the lethal condition of penetration by more than one sperm (Fauser *et al.*, 1999; Guraya, 2000a, b). During fertilization stored Ca^{2+} is mobilized raising the free Ca^{2+} level within the egg. This transient wave of Ca^{2+} stimulates the cortical granule reaction, a process similar to the acrosomal reaction. By fusion of the cortical granules with the overlying egg membrane, hydrolytic enzymes are released from the egg into the perivitelline space between the plasma membrane and zona pellucida. These enzymes change the ZP and especially its ZP_3 at the molecular level in such a way that it becomes impermeable for sperm: bound sperm detach and new sperm can no longer bind to the ZP (Fauser *et al.*, 1999; Wolf and Zelinski-Wooten, 2001). Niloson *et al.* (1999) have reported the expression of envelope proteins of endogenous C-type retrovirus on the surface of mouse and human oocytes at fertilization. Its expression enfeebled significantly after fertilization, suggesting its role at the sperm-egg binding and fusion.

Cortical and zona reactions in human oocytes and embryos have been studied by high-resolution analysis (Wolf and Lanzendorf, 1991). In unfertilized oocytes, cortical granules consisting of carbohydrates and proteins lie in the cortex in a uniform manner (Figs. 19, 20A, 20B) (Fauser *et al.*, 1999; Guraya, 2000a; Chapter 2). Following insemination, fertilized oocytes and embryos lose cortical granules, and cortical granule material finally lies in the perivetelline space (Figs. 20 (C-H)). Cortical granule contents get associated with the inner surface of the ZP as supported by the presence of striae of dense material at the site of interaction (Guraya, 2000b). In oocytes showing normal monospermic fertilization, sperm generally do not penetrate the inner ZP or within the perivitelline space. In polyspermic oocytes, however, supernumerary sperm can be observed penetrating the inner ZP and in the perivetelline space, possibly suggesting delayed cortical granule release (Wolf and Lanzendorf, 1991; Fauser *et al.*, 1999).

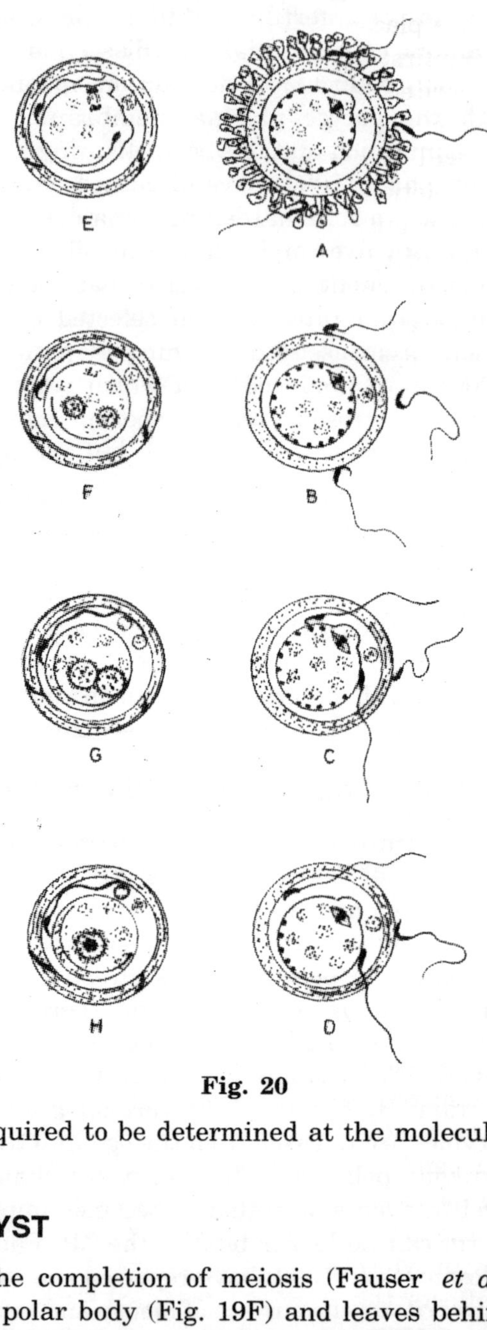

Fig. 20

Magerkurth *et al.* (1999) using scanning electron microscopy have demonstrated four different types of zona morphology, ranging from a porous, net-like structure to a nearly smooth and compact surface. No correlation could be established between zona type and oocyte maturity or zona type and achieved fertilization. However, fertilized polyploids oocytes had a more compact and smooth zona surface than unfertilized ones. But the number and distribution patterns of bound spermatozoa on the ZP showed extremely variable patterns regardless of the zona morphology. Wojcik *et al.* (2000) have suggested that paternal proteosomes of human spermatozoa enter the oocyte during fertilization in tight association with the centrioles and may serve a special function during further development, which can be associated with the function of hypothetical proteolysis center. This function is required to be determined at the molecular level.

5.8 FORMATION OF ZYGOTE AND BLASTOCYST

Fertilization activates the egg and triggers the completion of meiosis (Fauser *et al.*, 1999). This results in the extrusion of the second polar body (Fig. 19F) and leaves behind a haploid set of maternal chromosomes in the egg's female pronucleus. Meanwhile, the nuclear membrane of the sperm nucleus breaks down chromatin decondenses and is reorganised, and a new nuclear membrane is formed around the male pronucleus. Then the two pronuclei move towards each other (Fig. 19G). When they migrate, DNA replication takes place. Upon meeting, the two pronuclei do not fuse but their nuclear membranes break down, the chromosomes assemble on the metaphase plate and the first cell division takes place (Fauser *et al.*, 1999). Heikinheimo and Gibbons (1998) have suggested that the very first cell divisions of the human pre-embryo is still under the control of maternally inherited mRNA and protein as the destruction of C-mos and active M-phase promoting factor. Since the primordial oocytes undergo conspicuous changes during aging (Chapter 1) it will be rewarding to determine the effect of their aged ooplasm on the processes of completions of meiosis, extrusion of second polar body, breakdown of nuclear membrane of sperm nucleus, chromatin decondensation, formation of new nuclear membrane etc in the eggs of aged women as no information is available in this regard (Tevelde *et al.*, 2000; Bulletti *et al.*, 2001). In other words whether there occur chromosomal abnormalities or aberrations, which will certainly lead to developmental disorders during further embryogenesis.

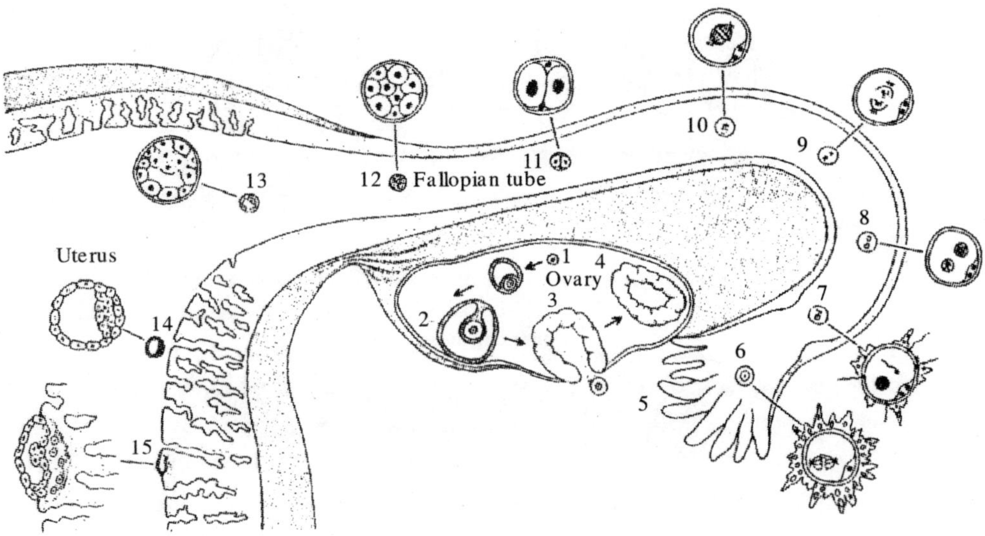

Fig. 21

The basic mechanism of blastocyst formation is similar in mammals including human. After compactation fluid accumulates between the intercellular spaces. (Feichtinger and Kemeter, 1987; Gossler, 1992; Wolf and Zelinski-Wooten, 2001). Outer cells pump fluid into the nascent cavity, which shows a rapid expansion. The timing of cavitation appears to depend on the nucleo-cytoplasmic ratio or DNA or chromosomal replication, but does not depend on the absolute number of cells or cell divisions in the zygote. When the number of cells in the embryo is experimentally decreased or increased or cell divisions where suppressed with cytochalasin B (which does not affect DNA replication) neither manipulation substantially affects the time of blastocyst formation. Two distinct cell populations are found in the blastocyst: an outer layer of trophectodermal (TE) cells, which constitute a true epithelium surrounding the fluid-filled blastocoel and the inner cell mass (ICM) cells, a group of cells which is attached to one side of the inner surface of the trophectoderm (Figs. 21 and 22). Allocation of cells either the trophectoderm or inner cells mass takes place during the late morula stage. Trophectoderm and inner cell mass appear to remain totally distinct lineages from the onset of cavitation (Feichtinger and Kemeter, 1987; Wolf and Lanzendorf, 1991; Fauser *et al.*, 1999; Wolf and Zelinski-Wooten, 2001). The trophectodermal cells give rise

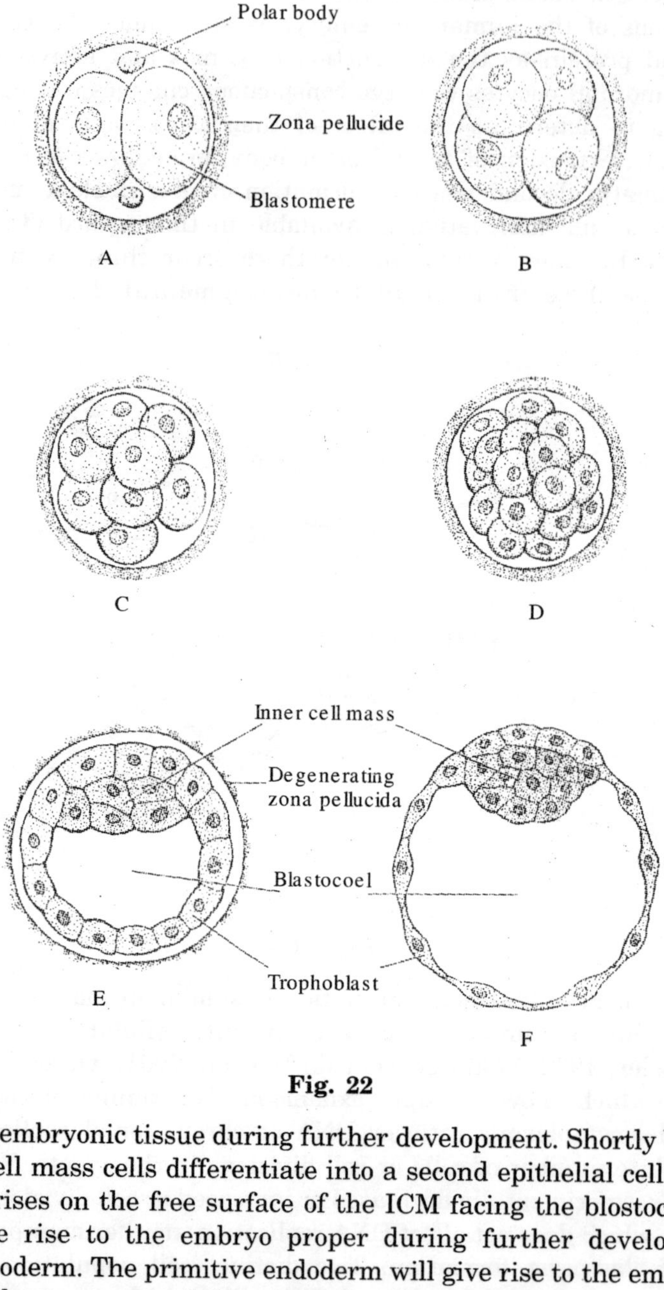

Fig. 22

exclusively to extra embryonic tissue during further development. Shortly before implantation some of the inner cell mass cells differentiate into a second epithelial cell type, the primitive endoderm, which arises on the free surface of the ICM facing the blostocoel. The remaining ICM cells will give rise to the embryo proper during further development and to the extraembryonic mesoderm. The primitive endoderm will give rise to the embryonic membranes i.e., the endodermal component of the visceral yolk sac and parietal yolk sac. It will be rewarding to determine the roles of growth factors at the cellular and molecular levels in the development and differentiation of ectoderm, mesoderm and endoderm and their subsequent roles in the formation of tissues and organs as recent studies have demonstrated the roles of cell to cell signals and growth factors during embryogenesis (Asashima *et al.*, 1999; De Laat *et al.*, 1989; Fauser *et al.*, 1999; Wolf and Zelinski-Wooten, 2001). It will also be interacting to make comparative cellular and molecular studies of blastocyst formation and the specific roles of trophectoderm and ICM cells in the aged women (Tevelde *et al.*, 2000; Bulletti *et al.*, 2001). The results of such studies will be helpful to determine if any abnormalities occur at this stage of development because of the effects of aging of primordial oocytes (as discussed in Chapter 1). If any abnormalities occur at the stage of blastocyst formation, *i.e.*, the formation of trophectoderm, inner cell mass ectoderm, mesoderm and endoderm of aged women, then these may lead to disorders in the formation of tissues and organs of infants. Therefore, such studies are very important for the assessment and the proper management of developmental disorders of infants with the help of techniques of genetic engineering, biotechnology etc (Wolf and Zelinski-Wooten, 2001). Vitrified human oocytes develop to the blastocyst stage with IVF (Hung *et al.*, 1999). A four-step thawing method (with 2.5 minutes interval) was more effective in supporting preimplantation embryo development than four-step thawing method with 5-minutes interval.

6

Early Development

With cleavage, the first differentiation step has already occurred (figs. 20, 21 (A-D)) and the allocation of cells either to embryonic or extraembryonic cell lineage has started. Two distinct cell populations, the inner cell mass and trophectoderm, are distinguishable (Figs. 21E and F). The embryo is still surrounded by zona pellucida, which is shed shortly before implantation (Fig. 20). During this stage the embryos are amenable to a variety of manipulations such as DNA transfer, chimera production etc (Wolf and Zelinski-Wooten, 2001). In the early stages of cleavage, the blastomers represent totipotent; equivalent cells as isolated single blastomers of two and four cells embryos can form blastocysts *in vitro* (Feichtinger and Kemeter, 1987). Blastocysts formed *in vitro* from isolated two cell embryos can develop into a normal human foetus after transfer into the uterus of the female (Bulletti *et al.*, 2001; Wolf and Zelinski-Wooten, 2001).

Fertilization as supported by the presence of male and female pronuclei, can be assessed between 12 and 18 hr after insemination of human oocytes *in vitro* (Feichtinger and Kemeter, 1987; Wolf and Lanzendorf, 1991). If permitted to continue in culture, such embryos can start cleavage and development to the blastocyst stage *in vitro* with differentiation of the trophectoderm, inner cell mass and primitive ectoderm as well as initial formation of germ layers. Based on pregnancy as the last-event, Cummins *et al.*, (1986) observed ideal growth (cleavage) rates for human embryos fertilized *in vitro* as the two-cell stage at 33.6 hr; the four-cell stage at 45.5 hr, and the eight-cell stage at 56.4 hr following insemination. Generally, the more rapid the cleavage rates, the more likely pregnancy will occur following transfer (Feichtinger and Kemeter, 1987; Wolf and Lanzendorf, 1991; Bulletti *et al.*, 2001; Wolf and Zelinski-Wooten, 2001). Additional criteria for the assessment of embryo quality include the size and shape of blastomeres, the presence of a nucleate fragment, and the appearance of the cytoplasm. The incidence of chromosomal abnormalities in embryos analyzed after IVF occurs in the range of 20-30% (Pleacht *et al.*, 1988; Plachot and Popseu, 1993; Bulletti *et al.*, 2001; Wolf and Zelinski-Wooten, 2001). A comparative study is required to be made of chromosomal abnormalities in embryos analyzed after IVF in young and aged women to determine the effects of aging of primordial oocytes (Tevelde *et al.*, 2000; Bulletti *et al.*, 2001) as discussed in Chapter 1.

Tesarik (1999) has reviewed the importance of calcium signaling in human preimplantation development with special emphasis on possible mechanisms by which inadequate or truncated calcium signals can impair embryo development. These mechanisms include complete failure of the second meiotic division, leading to triploidy; incomplete failure of the second meiotic division, resulting in *de novo* chromosomal numerical abnormalities; abnormal pronuclear development and function; abnormalities of the blastomere cell cycle possibly resulting in cleavage arrest, and problems with blastomere allocation to embryonic cell lineages, leading to disproportionate development of the inner cell mass and trophectoderm derivatives, which can be the origin of implantation failure or miscarriage (Fauser *et al.*, 1999; Bulletti *et al.*, 2001; Wolf and Zelinski-Wooten, 2001). Further investigations are required at the molecular level to decipher the nature of normal development signals, to determine the key check points at which these signals are required to inhibit the switch to apoptosis and to study the possibilities of the rapentie action at these checkpoints to rescue the endangered embryo for normal development.

6.1 MATERNAL AND ZYGOTIC REGULATION OF DEVELOPMENT

Early embryonic development is accompanied by qualitative alterations in protein synthetic activity originating in general, from the differential activation of stored mRNA as discussed in Chapter 2. Genomic participation as observed by qualitative alterations in protein synthesis and developmental sensitivity to the transcription inhibitor a-amanitin is usually associated with the four to eight-cell stage transition in the human (Braude *et al.*, 1988; Bulletti *et al.*, 2001; Wolf and Zelinski-Wooten, 2001). This information is relevant to recent research advances in the area of preimplantation diagnosis and in the treatment of genetic disorders in animal models (Monk *et al.*, 1987; Wilton *et al.*, 1989) and in primates. Zygote gene expression is initiated between the 4- and 8-cell stages after which the pre-embryo begins to utilize its own genes (Heikinoheime and Gibbons, 1998; Bulletti *et al.*, 2001; Wolf and Zelinski-Wooten, 2001). Some of the first genes to be expressed in the human pre-embryo encode proteins that are associated with cell division extracellular growth regulatory signals as well as with factors associated with implantation.

6.2 GENE ACTIVITY DURING CLEAVAGE OR EARLY DEVELOPMENT

Our knowledge is very meagre about the gene activity or regulatory proteins during cleavage development (Antezak Van Blerkom, 1990). However, a variety of alterations occur in the composition of RNA and proteins during cleavage and development in other mammalian species (Russo *et al.*, 1992; Gossler, 1992; Menezo and Ben-Khalifa, 1995) as well as in non-mammalian animal species (Russo *et al.*, 1992). But the rapidly increasing information obtained with other mammalian species in such domains as embryo freezing, embryo dissection, micromanipulation, and gene activity or insertion can not be adequately extrapolated to human embryos without the knowledge of similarities and differences between the mechanisms and timing of basic cellular and molecular events involved in the regulation of the process of early development and differentiation in man (Tevelde *et al.*, 2000; Bulletti *et al.*, 2001; Wolf and Zelinski-Wooten, 2001) and other mammals. It is well known that immediately after fertilization and for sometime during further development, the mammalian embryo is not yet able to utilize its own genetic message formed from the integration of the paternally and maternally derived genomes (Menezo and Ben-Khalifa, 1995; Tevelde *et al.*, 2000; Bulletti *et al.*, 2001; Wolf and Zelinski-Wooten, 2001). Developmental processes during this period, whose duration appears to vary remarkably among species, are completely under maternal (oocyte derived) genetic control which is exerted by maternal RNA molecules synthesized during oocyte growth and maturation (Chapters 2 and 3). The concentration of stored maternal species RNA decreased rapidly as total maternal RNA, polyadenylated RNA and specific messages for histone H_3 and actin are degraded. These appear to be only very little, if any, transcription from the embryonic genome during this time. Therefore, the transcriptional inhibitors like amanitin do not inhibit development of fertilized egg to the two-cell stage (Gossler, 1992; Menezo and Ben-Khalifa, 1995). Therefore, in the absence of embryonic gene activity, (Tesarik, 1987; Gossler, 1992; Menezo and Ben-Khalifa, 1995; Tevelde *et al.*, 2000; Bulletti *et al.*, 2001; Wolf and Zelinski-Wooten, 2001) as the morphologic cellular and molecular events occurring in early development are regulated by differential maternal messenger RNA (mRNA) activation, differential protein turnover and post translational modifications as demonstrated for the mouse (Gossler, 1992). RNA synthesis (embryonic gene transcription) may be absents at this time, but it may also take place. In the latter case, the embryonic genes would already be transcribed but the new transcripts would not function for immediate translation and phenotypical expression and thus remain stored for later use as it appears to be the case of mouse one-cell zygotes, which show a limited synthesis of RNA on the DNA template (Tesarik, 1987; Gossler, 1992; Menezo and Ben-Khalifa, 1995), but whose morphologic, cellular and molecular development changes are not apparently affected by physical enucleation or inhibition of transcription. From the reasoning as discussed here it can be stated that while describing the problem of embryonic gene activation, three largely independent processes must be kept in view: (1) activation of embryonic RNA synthesis, (2) the onset of translation of the new gene transcripts, and (3) the beginning of the embryonic gene expression. At the two-cell stage transcription from the embryonic genome begins and is needed for normal development to proceed as all classes of new RNA are synthesized and further development is blocked by inhibition of transcription with α-amanitin.

Newly synthesized proteins were found already upon fertilization by 2-D gel analysis of metabolically 35_S methionine labelled material from early embryos (Gossler, 1992; ; Menezo and Ben-Khalifa, 1995; Tevelde *et al.*, 2000; Bulletti *et al.*, 2001; Wolf and Zelinski-Wooten, 2001). The synthesis of these proteins does not need transcription as these are also synthesized when embryos are either incubated in the presence of the transcriptional inhibitor α-amanitin or mechanically enucleated, suggesting a post transcriptional regulation of this class of genes. The mRNAs coding for at least some of these proteins accumulate during oogenesis as discussed in Chapter 2. Following fertilization, translation from these mRNAs is greatly decreased (Chapter 5).

Proteins translated from newly synthesized RNA can be in two cell embryos using genetic variants of β-glucuronidase and β_2-microglobulin which are paternally encoded and can be distinguished from the maternally inherited form (Gossler, 1992; Menezo and Ben-Khalifa, 1995; Tevelde *et al.*, 2000; Bulletti *et al.*, 2001). As sperm dies not contribute RNA to the zygote, the presence of paternally encoded proteins establishes that newly transcribed embryonic RNA has been translated into protein. The transition from maternal to embryonic gene expression in the mouse occurs at the two-cell stage during its development and only the first cell division is exclusively under maternal control. Such cellular and molecular studies are required to be made from of gene expression during early embryogenesis of eggs from aged women (Tevelde *et al.*, 2000; Bulletti *et al.*, 2001; Wolf and Zelinski-Wooten,

2001). Transmission of altered DNA from smoking by spermatozoa has been demonstrated in preimplantation embryos and in association with increased risk of childhood cancer (Zenzes, 2000).

6.2.1 RNA Synthesis in Human Embryos

Incubation with 3H-uridine with subsequent detection of the incorporated radioactivity by autoradiography forms one of the conventional methods for the study of RNA synthesis in animal cells (Plachot and Popeseu, 1993). By using this method of RNA synthetic capacity of human preimplantation embryos obtained as supernumerary concept from IVF programs has been demonstrated (see Tesarik, 1987; Plachot and Popeseu, 1993; Tevelde *et al.*, 2000; Bulletti *et al.*, 2001; Wolf and Zelinski-Wooten, 2001). The appearance of RNA labelling in human four-cell embryos is believed to reflect the activation of RNA synthesis than a profound change in 3H-uridine uptake. By contrast to extranucleolar RNA synthesis, nucleolar RNA synthesis does not take place in human four cell embryos. Extranucleolar labeling provides a proof of 3H-uridine entry into the embryonic pool. Actually human two-to four-cell embryos do not show true nucleoli but merely their precussors (Tesarik, 1987; Tevelde *et al.*, 2000; Bulletti *et al.*, 2001). These nucleolus-like bodies (NLB) do not show any recently replicated embryonic DNA and do not show the fine structural features generally attributed to the machinery for RNA synthesis. The transformation of NLBs into active nucleoli involving DNA incorporation from adjacently placed chromatin and activation of nucleolar RNA synthesis starts after the third cleavage division. Thus, 3H-uridine incorporation into nucleoli can be found in some blastomeres of human six-cell embryos, and more frequently, in eight-cell embryos (Tesarik, 1987; Tevelde *et al.*, 2000; Bulletti *et al.*, 2001; Wolf and Zelinski-Wooten, 2001).

6.2.2 Alterations in Ribosomes and Other Cytoplasmic Components and their Implications in Development

As the ribosomal RNA (rRNA) is produced by the nucleoli, the alterations in nucleolar activity appear to account for changes in the amount of ribosomes produced during cleavage of human embryos *in vitro* (Tesarik, 1987; Tevelde *et al.*, 2000; Bulletti *et al.*, 2001; Wolf and Zelinski-Wooten, 2001). The ribosomal amount per embryo is decreased to less than a half between two-cell and four-cell stages. This decrease continues, though to a lesser degree, between four-cell and eight-cell stages. The overall decrease in the number of ribosomes is decelerated between the four-cell and eight-cell stages due to the production of new ribosomes in blastomers showing active nucleoli, whereas in the other blastomeres having no nucleolar activity the ribosomal amount is greatly decreased (Tesarik, 1987; Tevelde *et al.*, 2000; Bulletti *et al.*, 2001). These results show that the ribosomes are not formed during the first-three cycles of human preimplantation embryos, suggesting the utilization of only oocyte-inherited ribosomes for protein synthesis during this period. Thus, there occurs a progressive decrease in the number of maternally derived ribosomes in the absence of *de novo* formation of embryonic ribosomes that starts in eight-cell human embryos. Then the production of ribosomes occurs very fast from the eight-cell stage, as the newly produced ribosomes constitute a great amount of the ribosomal population.

The differential pattern of activation of RNA synthesis in blastomeres of human preimplantation embryos offers a new and possibly the only tool for the study of early signs of embryonic gene expression in humans (Tevelde *et al.*, 2000; Bulletti *et al.*, 2001). The activation of nucleolar RNA synthesis in some blastomeres of human eight-cell embryos does not form the only characteristic to distinguish these blastomeres from others as the activation of nucleolar transcription is always accompanied by a conspicuous increase in extra nucleolar RNA synthesis (Tesarik, 1987; Tevelde *et al.*, 2000; Bulletti *et al.*, 2001), indicating the presence of two different patterns of RNA synthesis in most human eight-cell embryos blastomeres, which are denoted as high RNA producers and low RNA producers by Tesarik (1987). The low-RNA-producing blastomeres (early cleavage pattern of RNA synthesis) show a low level of extra nucleolar RNA synthesis and the absence of nucleolar RNA synthesis. This pattern of RNA production is of common occurrence in blastomeres of human four-cell-embryos and in some blastomeres of eight-cell embryos and morulae. In contrast to this, the high RNA-producing blastomeres show very high concentrations of both extranucleolar and nucleolar RNA synthesis and can be seen only from the eight-cell stage onward. The increase in the amount of extranucleolar RNA synthesis in some blastomeres of human eight-cell embryos may be due to activation of transcription of new extranucleolar genes and, consequently, the newly produced RNA is partly the mRNA. The presence of low-RNA-producing and high-RNA-producing blastomeres in a single embryo provides a possibility for making a detailed morphologic comparison of the two kinds of blastomeres with the objective of finding and defining potential markers of early embryonic gene expression. Thus, providing a new basis for morphologic investigations of human embryos (Tevelde *et al.*, 2000; Bulletti *et al.*, 2001; Wolf and Zelinski-Wooten, 2001).

By using electron microscopic morphometry a quantitative comparison has been made of the occurrence of cytoplasmic structures between the low-RNA and and high -RNA-producing blastomeres of human eight-cell embryos in order to determine the precise expression of human embryonic genes as early as the eight-cell stage (Tesarik, 1987; Menezo and Renard, 1993; Bulletti *et al.*, 2001). The results obtained have suggested that there are two quantitative parameters volume density of tubular structures and of the Golgi apparatus for which the difference between the two types of blastomeres is highly significant ($P < 0.02$) and one parameter volume density of lysosomes for which this difference is significant ($P < 0.05$). The difference in the quantitative representation of tubules between the low-RNA-producing and high RNA-producing blastomeres of human eight-cell embryos and morulae is so conspicuous that it can be easily determined even by a simple examination of electron micrographs without using morphometric approach.

The various quantitative morphological differences in the occurrence of various cytoplasmic components as described above have been attributed either to expression of new RNA transcripts in the high-RNA-producing blastomeres, or to alterations occurring in the low-RNA-producing blastomeres and produced by the relative insufficiency of the proteosynthetic apparatus due to a decreased number of ribosomes (Tesarik, 1987; Menezo and Renard, 1993; Tevelde *et al.*, 2000; Bulletti *et al.*, 2001). The absence of statistically significant differences in any of the quantitative morphologic parameters studied between four-cell embryos and the low-RNA-producing blastomeres of eight-cell embryos suggests that the observed quantitative shift is a developmental change depending on the activation of embryonic gene transcription. Furthermore, if a defective function of the proteosynthetic machinery in the low-RNA-producing blastomeres were at the origin of the above mentioned quantitative, differences, insufficient renewal of cell organelles can be expected to be the immediate cause but this view is not supported by the data obtained (Tesarik, 1987; Menezo and Renard, 1993; Tevelde *et al.*, 2000; Bulletti *et al.*, 2001). Actually volume density of mitochondria occupying the second largest fraction of the cell volume among all cytoplasmic components investigated is even higher in the low-RNA-producing blastomeres of eight-cell embryos (as well as in four-cell embryos) than in the high-RNA-producing blastomeres, although this difference is not statistically significant. The significant enhancement in the amount of tubular structures in the high-RNA-producing blastomeres is associated with a decrease in vesicles, so appearing to be the result of differential activation of embryonic genes which regulate mechanisms controlling the pattern of endoplasmic reticulum membranes partition into vesicular and tubular structures. The decrease in the volume density of the Golgi Apparatus ensuring from the activation of RNA synthetic pattern in the high-RNA-producing blastomeres is believed to be a stage-specific developmental change controlled by a regulatory mechanism dependent on a specific part of the embryonic genome that is activated only in this type of blastomere (Tesarik, 1987). This embryonic message is believed to direct a redistribution of membranes among different cytoplasmic components without causing significant alterations in the existing steady state between the overall intracellular membrane degradation and production.

Besides the above quantitative morphologic markers of embryonic gene activity, three qualitative ultrastructural markers have been used. The first is the structural differentiation of nucleoli as already described. The other two qualitative morphologic signs of early human embryonic gene expression are the structural reorganization of the nuclear envelope with the disappearance of vesicular structures between the two leaflets and the first appearance of the granular endoplasmic reticulum, which shows its only a rare appearance in eight-cell embryos but is markedly increased in the subsequent morula and blastocyst stages. Both the quantitative and qualitative characteristics as discussed here show that human embryonic genes are expressed in the phenotype as early as eight-cell stage of preimplantation development, i.e., one cell cycle after the appearance of the first detectable embryonic gene transcription. But the expression of paternal genome in mouse embryos in which transcription is activated at the one-cell stage (Gossler, 1992; Menezo and Renard, 1993) can be demonstrated in the subsequent cell cycle, using biochemical and morphological markers (Tesarik, 1987; Tevelde *et al.*, 2000; Bulletti *et al.*, 2001). Phenotypical expression of the embryonic message transcribed one cell cycle ago appears to be a more general phenomenon in early mammalian embryogenesis, and it is also compatible with the results obtained using inhibitors of transcription. From the results of various studies as discussed here it can be stated that human embryonic gene transcription begins to occur from at least four-cell stage of preimplantation development. Embryonic RNA formation appears to be possibility that the uptake of 3H-uridine prior to the four-cell stage may not be so efficient at detectable level of incorporation into newly-produced RNA. But phenotypical expression of human embryonic genes can be detected at the eight-cell stage by quantitative ultrastructural analysis of blastomeres showing different patterns of RNA synthesis. But it remains to be determined whether there is any expression or translation of the newly transcribed embryonic message at the four cell stages. But if it occurs, it would be depending on the oocyte-inherited supply of ribosomes, as the production of embryonic ribosomes does not activate until eight-cell stage (Tevelde *et al.*, 2000; Bulletti *et al.*, 2001).

The results of various studies on the activation of the embryonic genome during early human embryogenesis as discussed above has some interesting implications as to the suitability of the oocyte for IVF and the chances of individual embryos grown *in vitro* of undergoing normal implantation and developing further after uterine replacement (Crozet, 1993; Guillomo *et al.*, 1993). In view of the delayed start of the formation of ribosomes in human embryos (Tesarik, 1987; Tevelde *et al.*, 2000; Bulletti *et al.*, 2001) as compared with the mouse embryo (Gossler, 1992; Menezo and Renard, 1993; Bulletti *et al.*, 2001), the amount of maternally inherited ribosomes, which are present in the human oocyte at the time of fertilization, constitutes one of the important facets of the oocyte developmental potential. Therefore, it will be rewarding to determine the variability of the ribosomal amount in human oocytes recovered for IVF. But there are some technical and ethical problems to carry out this type of research work (Wolf and Zelinski-Wooten, 2001) as the fate of each individual embryo that is transferred into the uterus is prior determined as early as before fertilization during the late preovulatory period when formation of new ribosomes is arrested (Guraya, 1985, 2000a, Chapter 2). Thus, ovum must meet its demands for protein stages synthesis during very early embryogenesis *i.e.*, upto the eight-cell stage from the prefabricated supply of ribosomes formed during oocyte growth. Some evidence is produced to show that the developmental potential of each human oocyte growing within large antral follicles may be strictly correlated with the pattern of RNA synthesis (Tesarik, 1987; Chapter 2). The number of ribosomes in the preovulatory human oocyte appears to be a major factor in its developmental potential (Chapter 2). Nonviable human embryos may develop seemingly normally to the morula stage (Tevelde *et al.*, 2000; Bulletti *et al.*, 2001). Early human development appears to be switched from the maternal to the embryonic control between the four and eight-cell stages (Tevelde *et al.*, 2000; Bulletti *et al.*, 2001). Thus, development is controlled solely by the oocyte-derived genetic information, and it is far from being sure for any respective embryo whether it will adequately activate its genome and survive or whether it will not do so and will die. The definition of the qualitative and quantitative morphologic markers of embryonic genome expression in human conception as discussed above provides a possibility of a physiologically founded interpretation of ultrastructural data for making adequate evaluation of embryo and blastomere viability. If the newly produced extranuclear RNA (possibly at least in part mRNA) at the four cell stage should carry information whose immediate expression would be required to support further normal development but this message could only be translated on maternally derived ribosomes. In regard to the sudden decrease in the number of ribosomes between the four and eight-cell stage possibly represents a vulnerable period in the early human embryogenesis when it will be decided whether the embryo will survive or die. But the results of this type of situation cannot be demonstrated with conventional criteria of embryonic quality for making a selection of viable embryo to be transferred. Actually, human embryos, which fail to undergo the overall acceleration of RNA production and are most probably destied to die can develop further, developing normal compacted morulae and lacking unequivocal ultrastructural abnormalities (Tesarik, 1987; Tevelde *et al.*, 2000; Bulletti *et al.*, 2001). Plachot *et al.*, (1987) after making morphologic and ultrastructural study of human embryos obtained by *in vitro* fertilization have suggested that the morphologic aspect of the embryos does not correlate with their structure or developmental capacity except for group of 4 degenerated embryos (Plachot and Popeseu, 1993). Multinucleated blastomeres appear to be the most important abnormality. All such supernumerary nuclei show an active DNA synthesis pattern, but can be synchronized in their mitotic activity. Only 42% of the embryos are viable. Preliminary experiments with human oocytes fertilized *in vitro* but rejected for ET have shown that chromosomal and biochemical studies could be carried out prior to implantation (Verlinsky *et al.*, 1987; Menezo and Ben-Khalifa, 1995). But such an approach raises considerable ethical concerns, especially those dealing with the implications of cloning and sex selection (Tevelde *et al.*, 2000; Bulletti *et al.*, 2001; Wolf and Zelinski-Wooten, 2001).

6.2.3 Chromosomal Disorders (Abnormalities) and their Implications in Development

The origin of human chromosomal disorders is of fundamental importance to *in vitro* fertilization (IVF) (Plachot and Popeseu, 1993; Fauser et al.,1999; Tevelde *et al.*, 2000; Bulletti *et al.*, 2001; Wolf and Zelinski-Wooten, 2001). Routinely gametes and embryos are handled in IVF laboratories, and those hold the secrets of causes of human trisomies, monosomies and polyploidies (Plachot and Popeseu, 1993; Tevelde *et al.*, 2000; Bulletti *et al.*, 2001; Wolf and Zelinski-Wooten, 2001). Thus, chromosome and gene anomalies, are in man the principal cause of embryonic and neonatal death and various infant disorders. Indeed, if 0.6% of living new borns carry a chromosomal disorder, this evidence is ten times higher in stillbirths and 100 times in spontaneous miscarriages (Plachot and Popeseu, 1993; Bulletti *et al.*, 2001). Moreover, nearly one infant out of 100 is afflicted at birth by one of the 4,000 odd Mendelian diseases known today. But the causes of these are required to be determined at the molecular level during aging of primordial oocytes and subsequent oocyte growth and maturation. However, most of these conditions induce errors of organogenesis responsible for more or less severe malformations during development of embryo: lastly they may interfere with human reproduction either by acting directly on gonadal development or indirectly on gametogenesis (Guraya, 1998a; Vats 1999; Sharma and Vats, 2002).

Numerical anomalies are due essentially to meiotic non-disjunction, and result in trisomes (presence of a normal super numerary chromosome) or monosomies (loss of a chromosome). Autosomal trisomy forms a serious conditions, resulting in the birth of children with mental retardation (Edwards, 1987; Fauser et al.,1999; Tevelde *et al.*, 2000; Bulletti *et al.*, 2001; Wolf and Zelinski-Wooten, 2001). Chromosomal imbalance in embryo forms a characteristic feature of human pregnancy, with some commentators claiming that half of all human conceptions are chromosomally imbalanced, (Plachot and Popeseu, 1993; Tevelde *et al.*, 2000; Bulletti *et al.*, 2001). The most common anomalies are the autosomal and sex-linked trisomies and monosomies. A few triploid children (mostly mosaics) were born carriers of severe multiple, malformations responsible for neonatal deaths (Plachot and Popeseu, 1993; Tevelde *et al.*, 2000; Bulletti *et al.*, 2001).

Most autosomal trisomes originate in the ovary as demonstrated in studies using chromosome banding, and highest frequently is associated with the first meiotic division of the oocyte (Plachot and Popeseu, 1993; Fauser et al.,1999; Tevelde *et al.*, 2000; Bulletti *et al.*, 2001; Wolf and Zelinski-Wooten, 2001). The most important approach is to study the causes of chromosomal disjunctions in the first meiotic division of the oocyte at the molecular level. Two theories are put forth to explain the high incidence of autosomal trisomies originating from the meiotic division of the oocyte-one suggests that anomalies originate during ovulation, or at fertilization and chromosomes segregate abnormally as a result. Delayed ovulation disorders in follicle growth and after have all been raised as possible causes (Edwards, 1986; Plachot and Popeseu, 1993; Tevelde *et al.*, 2000; Bulletti *et al.*,

2001). The second theory, owned by Ewards and coworkers is totally different stating that chromosomal pairing failed when the eggs were formed in the foetal ovary (Guraya, 1998a), so that they segregate abnormally into the first polar body. Such abnormal oocytes are formed late in the foetus and conserved in older ages in the mother, so explaining the high incidence of these anomalies in older mothers (Plachot and Popeseu, 1993; Tevelde *et al.*, 2000; Bulletti *et al.*, 2001). Guraya (1999a) has suggested that aging or decay of primordial oocytes in aged women may affect meiotic non-disjunction leading to an increased incidence of chromosome anomalies. IVF has, however, allowed confirmation of the role of maternal age in the occurrence of chromosome abnormalities in the oocyte, since the incidence of aneuploidy increases from 24% in patients under 35 years to 38% in patients 35 to 42 years old (Tevelde *et al.*, 2000; Bulletti *et al.*, 2001; Wolf and Zelinski-Wooten, 2001). These differences may be due to aging of primordial oocytes at the cellular and molecular levels as discussed in Chapter 1.

Plachot and Popeseu (1993) have stated that occurrence either *in vivo* and *in vitro* gamete aging is detrimental to fertilization and embryo development through an increased incidence of chromosome imbalance (Tevelde *et al.*, 2000; Bulletti *et al.*, 2001). These abnormalities could be the consequence of modifications of microtubule organization resulting in the constitution of abnormal spindles and consequently to chromosome spreading leading to aneuploidy concerning groups of chromosomes. The factors involved in modifications of microtubule and mechanisms of their actions in aged oocytes/eggs are required to be determined at the molecular level. However, according to both the severity of the anomaly and the genetic content of the chromosomes involved chromosome aberrations have various consequences for embryo development, either, before implantation (embryo degeneration; cleavage arrest) or after implantation (spontaneous abortions) or at birth (stilborns, malformations) when analyzed at day I after insemination (Tevelde *et al.*, 2000; Bulletti *et al.*, 2001), two main abnormalities, either coincident with or directly related to fertilization have been reported, attested by an abnormal number of pronuclei, parthenogenetic activation (I pronucleus) or triploid (3 pronuclei). The capacity for cleavage of either chromosomally normal or abnormal zygotes is found to be slight different. The development and survival of embryo depends on the type of anomaly (Plachot and Popeseu, 1993; Tevelde *et al.*, 2000; Bulletti *et al.*, 2001). The developmental arrest occurs sooner when the incidence of chromosome anomalies is the more elevated (Wolf and Zelinski-Wooten, 2001).

Studies are required to be made of the first meiotic division of the oocyte to analyze chromosomal pairing and also of embryos to identify the origin of extra chromosomes. Such studies can be carried out in clinics practissing IVF, where a good supply of oocytes maturing *in vivo* or *in vitro* is available (Crozet, 1993; Tevelde *et al.*, 2000; Bulletti *et al.*, 2001; Wolf and Zelinski-Wooten, 2001). Such investigations are important for their own interest, as it is necessary to know how many embryos growing *in vitro* are abnormal and the known causes of those abnormalities are not some disorders of children which can arise through fertilization involving abnormal spermatozoa such as diploid spermatozoon, or rarely, the inclusion of the second polar body in the oocyte. (Plachot and Popeseu, 1993; Tevelde *et al.*, 2000; Bulletti *et al.*, 2001; Wolf and Zelinski-Wooten, 2001). Studies on this point would be very rewarding in the alleviation of abortion and birth anomalies both after conception *in vivo* and *in vitro*.

6.3 VARIOUS TYPES OF INFANT DISORDERS

During the recent years, various types of infant disorders pertaining to anatomy or (morphology, physiology, diabetes, allergy, behaviour (or nervous), hearing, vascularity, nephropathy, pathology, biochemistry, immunology, molecular biology, cytogenetics, stress etc. have been reported. Various types of infant disorders include leukocyte adhesion deficiency disorder (Hung *et al.*, 1999), intractable wheezing and swallowing problem in an infant (Chen and Huang, 1999), feeding high-risk infants with family history and allergy (Contencin, 1999), conflict between mothers with eating disorders and their infants during mealtimes (Stein *et al.*, 1999), galactosaemia (Murila, 1999); alveolar capillary dysphasia (Hintz *et al.*, 1999); cerebral blood flow velocities with *moyamoya* disease (Ipsiroglu *et al.*, 1999), infantile spasm (Baram *et al.*, 1999), patterns of pregnancy loss, perinatal mortality, and postneonatal childhood deaths in families of girls with Rett syndrome (Fyfe *et al.*, 1999), neonatal diabetes mellitus with hypergalactosemia (Kentrup *et al.*, 1999), phenylketonuria (Koch, 1999), hearing-impairment (Pickett and Ahlstrom, 1999), sulfite oxidase deficiency (Edwards *et al.*, 1999), hydrocephalus associated with glycogen storage disease type II (Sahin and du Plessis, 1999), deficient α–β 4 integrin expression in the gut associated with intractable diarrhea, (Lachaux *et al.*, 1999), carbohydrate-deficient glycoprotein syndrome in a new born with an unbalanced chromosomal translocation (Bowling *et al.*, 1999), prenatal diagnosis of thoracopelvic dysphasia (Hsieh *et al.*, 1999), Cavenous hemangioma of the skull in a neonate (Yoshida *et al.*, 1999), autosomal dominant inheritance of multicystic dysplastic kidney (Srivastava *et al.*, 1999) lipoamide dehydrogenase due to a novel mutation in interface domain (Shany *et al.*, 1999), neuronal pathogenesis of aspartylglucosaminuria in relation to expression of aspartylglucosaminidase in brain during development (Unsitalo *et al.*, 1999), nasal fosse dimensions in the neonate and young infant (Contencin *et al.*, 1999), small for gestational age infant in association with maternal prothrombin gene variant (Verspyck *et al.*, 1999), auditory neuropathy (Rance *et al.*, 1999), severe hypertrophic cardiomyopathy in Pompe's disease (Metzl *et al.*, 1999), pulmonary hemorrhage among infants (Dearborn *et al.*, 1999), cardiac rhabdomyomas (Smith and Sperling, 1999), early postnatal depression mood (Bergant *et al.*, 1999), pyruvate carboxylase deficiency (Brun *et al.*, 1999), lamellar ichthyosis (Tok *et al.*, 1999), intracranial haemorrhage (Totan and Albayrak, 1999), reye's syndrome (Belay *et al.*, 1999), absence of cysteinyl leukotriences in cerebrospinal fluid (Mayatepek *et al.*, 1999), isovaleric acidemia with promyelocytic myeloproliferative syndrome (Gilbert-Barness and Barness, 1999), clinical spectrum of infantile free sialic acid storage disease (Leymyre *et al.*, 1999), neuropathy in cobalamin-deficient breast-fed infants of vegetarian mothers (Renault *et al.*, 1999), short-chain hydroxacyl-coenzyme A dehydrogenase deficiency (Treacy *et al.*, 2000), prolonged activation of hypothalamus-pituitary-gonadal axis in a child with x-linked adrenal hypoplasia congenita (Takahashi *et al.*, 2000), neonatal alloimmune neutropenia in premature monozygous twins (Felix and Calboun, 2000), cytochrome oxidase deficiency presenting at birth as phyxia (Wills *et al.*, 2000), cranial MRI in neonatal hypernatraemic dehydration (Korkmazo *et al.*, 2000), mucopolysaccharidosis type IIB (Hunter syndrome, complicated by autoimmune hemolytic anemia (Mullen *et al.*, 2000), serum valproate levels in breastfeeding mother-infant pairs (Piontek *et al.*, 2000), affective communication of infants with autistic spectrum disorder and internal representation of their mothers (Kobayashi, 2000), the stressed neonatal kidney or neonatal vasomoto nephropathy (Toth-Heyn *et al.*, 2000), possible traumatic stress disorder in an infant with cancer (Roy and

Russell, 2000), hydrocephalus and the reproductive health of women: the medical implications of maternal shunt dependency in 70 women and 138 pregnancies and the proper management of these patients can lead to normal pregnancy and delivery (Liakoss *et al.*, 2000).

The precise causes of these infant disorders are required to be determined experimentally at the cellular, molecular, cytogenetical levels etc. in relation to the age and health of mothers (Tevelde *et al.*, 2000; Bulletti *et al.*, 2001). The effects of drugs, chemicals, food etc consumed by mothers as well as of xenobiotic factors during fertilization and early development also need to be studied or must be kept in mind. The incidence of intrauterine growth retardation on (IUGR) and foetal anomalies are generally higher in elderly mothers as compared to younger mothers (Tevelde *et al.*, 2000). Congenital anomalies and IUGR are common in children born to elderly women (Vats 1999; Sharma and Vats, 2002). The causes of these anomalies are required to be determined in relation to the age of the mothers at the cellular and molecular level as eggs especially primordial oocytes are deteriorated is quality as the women have aged, (Chapter 1), presenting a major obstacle to pregnancy and possibly to the normal development of embryo. In late thirties, the *in vitro* fertilization has been used and women get pregnant, but they are found to miscarry probably due to the quality of egg which has received much attention at cellular and molecular level to determine the effects of aging of women on the quality of eggs (Tevelde *et al.*, 2000; Bulletti *et al.*, 2001; Wolf and Zelinski-Wooten, 2001). It is now commonly believed that the quality of eggs is affected with aging of women (Introduction). Reproductive tract infections, drugs, smoking, tobacco chewing, anemia, low weight, malnourishment, cardiac diseases, nephritis, hypertension etc. in the mother can also lead to IUGR (Vats 1999; Sharma and Vats, 2002).

7

Biomedical and Clinical Implications of Aging Changes in Oocytes

The results of various histochemical and electron microscopic studies on human primordial follicles (or oocytes) as correlated and discussed here in the light of recent advances in cell and molecular biology have indicated that they undergo aging changes as evidenced by conspicuous alterations in some of their ooplasmic components during the later years (*i.e.*, third and fourth decades) of reproductive life of women (Chapter 1); which have not received any attention in the recent books (Fauser *et al.*, 1999; Tevelde *et al.*, 2000; Bulletti *et al.*, 2001). The ooplasmic components such as cytocentrum, spherical bodies, annulate lamellae, ribosomes, polysomes etc. consist mainly of informational molecules such as various RNAs (mRNA, rRNA and tRNA) and proteins; RNAs are well known to be produced by the differential activity of various maternal genes during the prophase of meiosis and stored in the ooplasm (Bachvarova, 1985; Guraya, 1998a; Fauser *et al.*, 1999; Tevelde *et al.*, 2000). The transcription of these genes is also supported by the autoradiographic studies of decondensed chromosomes in the nucleus of primordial oocytes (Guraya, 1974, 1998a) as already stated in Chapter 1. The aging changes in various informational macromolecules especially RNAs, proteins and DNA of spherical bodies, mitochondria, granular basophilic substance (consisting of ribonucleoproteins, and some lipoproteins constituting elements of ER and annulate lamellae) and possibly other organelles of maternal origin are expected to occur in terms of somatic cell life (Harford, 1995) as some primordial follicles continue to remain quiescent in human ovaries for 40 years or more, depending upon the individual variations in the initiation of menopause (El-Badrawi and Hafez, 1980; Gindoff and Jewelewicz, 1986; Fauser *et al.*, 1999; Tevelde *et al.*, 2000; Bulletti *et al.*, 2001).

The aging changes of ooplasmic components in primordial oocytes must be influencing the developmental processes of oocyte growth and maturation in the follicles getting ready for ovulation as the qualitative and quantitative changes in protein synthetic activity, which take place during oocyte growth and maturation and progression of the zygote, are largely if not solely dependent upon maternal informational macromolecules accumulated in the ooplasm (Bachvarova, 1985; Van Blerkom, 1989; Fauser *et al.*, 1999; Tevelde *et al.*, 2000; Bulletti *et al.*, 2001; Wolf and Zelinski-Wooten, 2001 Chapters, 2 and 3). These molecular changes in various components of primordial follicles during later years (i.e., third and fourth decades) of reproductive life of women form an important area of experimentation that could provide evidence to explain various types of disorders during oogenesis including ovum maturation and ovulation, fertilization and embryogenesis *in vivo* and *in vitro* during the later years of reproduction life; no attention has been paid in this regard (Fauser *et al.*, 1999; Tevelde *et al.*, 2000; Bulletti *et al.*, 2001; Wolf and Zelinski-Wooten, 2001). A deeper understanding of alterations in informational macromolecules formed during fertilization and early embryogenesis are hypothesized, which are required to be extended and confirmed with modern molecular probes (Fauser *et al.*, 1999). It is also emphasized that attempt must be made to correlate the incidences of various types of disorders in development or anatomy growth, maturation, behaviour or psychology etc. of children to the age of their relatively older mothers for developing better biomedical and clinical strategies to overcome for various types of infant disorders as described in detail in Section 6.3 of Chapter 6. Very little attempt has been made to correlate these various types of infant disorders to the age of mothers (Tevelde *et al.*, 2000; Bulletti *et al.*, 2001). Tangrray (2000) after giving a 10-year review of pervasive developmental disorders in human has concluded that private and government agencies must continue to support basic and applied research with special emphasis on genetic and molecular factors, the role of xenobiotic factors, chemicals, drugs, radiation, temperature etc. on developmental processes of human embryos leading to infant disorders which cannot be ignored and thus studies in this regard are required to be carried out at the molecular level (Tevelde *et al.*, 2000; Bulletti *et al.*, 2001; Wolf and Zelinski-Wooten, 2001; Vats 1999; Sharma and Vats, 2002).

Practically nothing is known about influences of aging changes in informational macromolecules of primordial oocytes on infant disorders at the molecular level (Tevelde *et al.*, 2000; Bulletti *et al.*, 2001; Wolf and Zelinski-Wooten, 2001) and thus investigations in this regard are required to be initiated. However, Battaglia *et al.* (1996) have observed the adverse effects of maternal age on meiotic spindle assembly in oocytes from naturally cycling women which may be the result of decay in molecular regulatory mechanisms of aged or devitalized oocytes. Tarin (1996) after discussing the aetology of age-associated aneuploidy has proposed a mechanism based on the free radical theory of aging. Hassold *et al.* (1996) have discussed the recent molecular data on the mechanisms of origin of different aneuploid conditions, the basis of maternal age in aneuploidy and the importance of aberrant genetic recombinations on the genesis of aneuploidy (Plachot, 1995; Homburg and Shelef, 1995; Tevelde *et al.*, 2000; Bulletti *et al.*, 2001). The development of molecular probes (tagged antibodies, labelled RNAs) to identify proteins whose expression or activity (*i.e.*, function) is affected by aging changes of primordial follicles as well as by environmental factors will go a long way in providing the analytical ability to test this hypothesis. Actually to get further insight into the etiology of aneuploidy in female germ cells, information is required about the chemical interactions between endogenous and exogenous compounds and those involved with oocyte meiotic maturation (de Laat *et al.*, 1989; Mailhes, 1995; Fauser *et al.*, 1999; Tevelde *et al.*, 2000; Bulletti *et al.*, 2001).

Any qualitative and quantitative alterations in the ooplasmic components at the molecular level in the oocytes of older mothers are expected to influence the development, growth and differentiation of embryos by sending abnormal stimulatory and inhibitory signals from the aged or devitalized ooplasmic components or decayed informational molecules to the genes present in the nucleus as the cytoplasmic factors are now known to regulate the differential activity of genes in eukaryotic cells (de Laat *et al.*, 1989; Papavassiliou, 1996; Nigg, 1997; Fauser *et al.*,1999). Actually transcription factors are central to the process of eukaryotic gene control. These proteins are known to influence the basal level of gene expression in a cell and can modulate genetic programs by activating or repressing the transcription of particular genes in a cell-type specific or inducible manner. Practically nothing is known about the molecular aspects of these signals and transcription factors produced by the aged or devitalized ooplasmic components of primordial oocytes in older mothers which are required to be investigated with modern immunocytochemical and molecular probes *in vivo* and *in vitro* (Tevelde *et al.*, 2000; Bulletti *et al.*, 2001; Wolf and Zelinski-Wooten, 2001), although cell to cell signals in mammalian development are discussed by various workers (de Laat *et al.*, 1989).

The chromosomal abnormalities occurring during human oogenesis are receiving an increasing attention at the molecular level to determine their effects on various types of human disorders during embryogenesis *in vitro* systems (Van Blerkom, 1989; Hassold *et al.*, 1996; Fauser *et al.*, 1999; Tevelde *et al.*, 2000; Bulletti *et al.*, 2001). Both before and after dictyate stage of meiosis the oocyte has less repair capacity and/or is more sensitive to DNA damaging agents. Epigenetic factors associated with the expression of genetic faults arising in oocytes have been largely ignored in the past and deserve immediate attention for the understanding of capacity of oocytes to repair genetic damage (Ashwood-Smith and Edwards, 1996; Fauser *et al.*, 1999; Tevelde *et al.*, 2000; Bulletti *et al.*, 2001; Wolf and Zelinski-Wooten, 2001). As early as 15 years before menopause, a woman's egg production starts to decline and more of the eggs she does produce contain chromosomal problems that make infertility, miscarriage, and birth defects more likely. This is due to the fact that working women delay child rearing as these days in some countries especially industrialized there is inadequate social safety net for working mothers.

Although numerous studies have been made of human fertilization and early embryogenesis *in vitro* systems all over the world (Trounson, 1985; Feichtinger and Kemeter, 1987; Van Blerkom, 1989; Snell and White 1996; Jennings *et al.*, 1996; Hiroi, 1996; Trounson and Bongsou, 1996; Conway-Myers and Steinkampf, 1996; Pulasz *et al.*, 1996; Gomel and Leung, 1997; Fauser *et al.*,1999; Tevelde *et al.*, 2000; Bulletti *et al.*, 2001; Wolf and Zelinski-Wooten, 2001; chapters 5 and 6), very little or no attempt has been made to study the effects of aging of primordial oocytes on *in vitro* fertilization and early embryogenesis taking into consideration the age of donors. However, there is observed an age-dependent decrease in embryo implantation after *in vitro* fertilization (Van Kooij *et al.*, 1996; Tevelde *et al.*, 2000; Wolf and Zelinski-Wooten, 2001). Yaron *et al.* (1995) have also made some studies of *in vitro* fertilization and oocyte donation in women 45 years of age and older of the 52 standard IVF cycles oocytes were retrieved successfully in only 32, of these, fertilization and embryo transfer were performed in 21 cycles. None of these treatment cycles resulted in clinical pregnancy. Age-dependent decrease in embryo implantation in all these cases can be attributed to the age-related decrease in embryo quality after the age of 37 years as the results of a recent study have suggested that endometrial receptivity is unaltered by age or diagnosis (Paulson *et al.*, 1997; Tevelde *et al.*, 2000). With advanced age of women, the perinatal complications have been related to their higher incidence of multiple gestation (Wolff *et al.*, 1997).

Besides the accelerated increase in the incidence of chromosomally imblanced after the age of 37 years, oocyte quality may also be affected by more gradual processes in the oocytes (Keefe *et al.*, 1995; Homburg and Shelef, 1995) as hypothesized here by the present author. The developmental abnormalities and the demise of the conceptus prior to the stage of implantation may stem from the poor quality of the oocyte (Homburg and Shelef, 19955; Hull, 1995), especially of various cytoplasmic components including, organelles, various RNAs, enzymes, proteins etc. The chromosomal abnormalities occurring during the maturation of human oocytes obtained from older women as well as due to the effects of various chemical and physical factors may be the result of abnormal molecular signals produced by the aged or decayed or damaged ooplasmic macromolecules, which form promising area for future investigations. Actually development failure or aberration may be the consequence of a prior perturbation in the normal sequence of chromosomal maturation and molecular and cytoplasmic differentiation of the oocyte, which is required to be investigated at the molecular level (Tevelde *et al.*, 2000; Bulletti *et al.*, 2001).

In various clinics of the world, very little or no attempt has been made to correlate the incidences of various types of human disorders in development, growth, maturation, behaviour, physiology etc. of children born to relatively older mothers (Section 6.3 of Chapter 6). In my own opinion, such biomedical and clinical studies are required to be made for the better understanding of effects of aging of primordial oocytes (Tevelde *et al.*, 2000; Wolf and Zelinski-Wooten, 2001) as well as effects of various chemical and physical factors on development of infants with significant birth defects (Vats, 1999; Sharma and Vats, 2002).The causes of developmental failure and various types of abnormalities during human embryogenesis may also be due to inherent defects as a result of aging of primordial follicles. Molecular probes that differentially resolve changes in various components of growing oocytes derived from such aged primordial follicles can be applied to the study of their differentiation (Fauser *et al.*, 1999). The rapid progress in biological and biomedical sciences in the last twenty years has brought with it an extensive development of the methods of molecular genetics and biology, which will be very helpful to reveal the precise nature of molecular changes in primordial follicles (oocytes) and their subsequent effects on developmental processes during oocyte maturation and embryogenesis. The various incidences of clinical disorders affecting semen quality (Jequier, 1995) may be the aging effects of primordial oocytes in the ovaries of their older mothers, which have not received any attention (Tevelde *et al.*, 2000; Bulletti *et al.*, 2001).

Environmental factors may also be playing some role in this regard. But further studies are required to approve or disapprove this hypothesis by recording the age of mothers in various programmes of assisted human reproduction which are being carried out extensively in clinics all over the world (Tevelde *et al.*, 2000; Wolf and Zelinski-Wooten, 2001). Very little or no attention is being paid to this aspect of human disorders in physiological processes (Tanguay, 2000). This author after completing a 10-year review of pervasive developmental disorders in human has recommended that private and government agencies must continue to support basic and applied research on problems of developmental disorders which are reported in detail in Section 6.3 of Chapter 6. The results of such biomedical and clinical studies will be very rewarding to advise the older mothers not to conceive during the later years of reproductive life and thus will have impact on society in many fields including family planning programmes from a scientific and philosophical point of view as well as from a cultural and social perspective. It will be interesting to mention here that the campaign, sponsored by the American Society of Medicine, actually addresses several risk factors for infertility including smoking, working too little or too much and having a history of sexually transmitted infections. But the age message has gotten all of the attention as women past their mid-30s are quite often unable to have a baby-and more at risk of problem pregnancy.

Zenzes (2000) has reported the effects of smoking on reproduction in relation to gene damage to human gametes and embryos. Assisted conception is a useful methodology for detecting disturbance in clinical outcome, meiotic maturation, and genetic integrity of human gametes in regard to effects of drugs, xenobiotic factors, smoking etc. Germinal cells are vulnerable to genetic damage from smoking, but can repair damage during meiosis. But in ejaculated spermatozoa, repair capacity declines drastically (Zenzes, 2000). Smoking changes the meiotic spindle of oocytes and spermatozoa, resulting in chromosome errors, which affect reproductive outcomes. Smoking is associated with reduced number of retrieved oocytes, leading to early age of menopause and is also associated with various reproductive disorders in women and men (Vats, 1999; Sharma and Vats, 2002).

References

Aaaitoonen, J., Laitinen, M.P., Vuojolainen, K., Jaatinen, R., Horelli-Kuitunen, N., Seppa, L., Louhio, H., Tuun, T., Sjoberg. J., Butzon, R., Hovata, O. Dale, L, and Kitvos, O. (1999). J. Clin. Endocrinol. Metab. 84, 2744.

Abir, R., Roizman, P., Fisch, B., Nitke, S., Oken, E., Orviueto, R., and Ben Rafael, Z. (1999). Hum. Reprod. 14, 1299.

Abhilasha and Krishna, A. (1996). J. Reprod. Fertil. 106, 202.

Aboulghar, M., Evers, J.H., and Al-Inany, H. (2000). Cochvaue Database Syst. Rev CD 001302

Adashi, E.Y. (1990). In: Gamete Physiology, R.H. Ash, J.P. Balmaceda and J Johnoton (Eds.). Serono Symposia, USA, Norwell, pp. 187-196.

Afzulpurkar, A. and Gupta, S.K. (1997). Amer. J. Reprod. Immunol. 38, 26.

Afzulpurkar, A., Shibahara, H., Hasegawa, A., Koyama, K., and Gupta, S.K. (1997). Hum. Reprod. 12, 2664.

Ahmadi, A. and Ng, S.C. (1999). Hum. Reprod. 14, 2279.

Ahmed, T., Gupta, B.C., Sidhu, K.S., and Guraya, S.S. (1984). Indian J. Anim. Res. 5, 15.

Albertini, D.F., Allworth, A.E., and Messinger, S.M. (1993). In. Ovarian Cell Interactions, Genes to Physiology, A.W. Hsueh and D.W. Schomberg (Eds.). Serono Symposia, USA, Springer Verlag, New York, 79-88.

Alexander, N.J., Griffin, D., Spierler, J.M., and Waites, G.M.H (Eds.) (1989). Gamete Interaction. Prospects for Immunocontraception. Wiley-Liss, New York.

Amsterdam, A., Keren-Tal, I., and Aharom., D. (1996). Steroids, 61, 252.

Anderiesz, C., Fong, C.V., Bongse, A., and Trounson, A.O. (2000a). Hum. Reprod. 15, 379.

Anderiesz, C., Ferraretti, A., Magli, C. Fiorentino, A., Fertini, D., Gianaroli, I., Jones, G.M., and Trounson, A.O. (2000b). Hum. Reprod. 15, 1140.

Arakl, M., Fukumatsu, V., Katabuchi, H., Shuttz, L.D., Takahashi, K., and Okamura, H. (1996). Biol. Reprod. 54, 478.

Anserini, P., Casta, M., Remorgida, V., and Venturini, P.L. (2000). Gynecol. Endocrinol. 14, 75.

Antezak, M. and Van Blerkom, J. (1999). Hum. Reprod. 14, 429.

Asashima, M., Kinoshita, K., Arizomi, T., and Malacinski, G.M. (1999). Inter. Rev. Cytol. 191, 2.

Ashwood-Smith, M.J., and Edwards, R.G. (1996). Mol. Hum. Reprod. 2, 46.

Aslam, I. and Fishel, S. (1999). Hum. Reprod. 14, 1528.

Baca, M. and Zamboni, Z. (1967). J. Ultrastruct. Res. 19, 354.

Bachrova, R. (1985). In: Developmental Biology. A Comprehensive Synthesis. Vol. 1. Oogenesis. I.W. Browder (Ed.), Plenum Press, New York, pp. 453-524.

Bachvarova, R. (1988). In: Meiotic Inhibition, Molecular Control of Meiosis: Progress in Clinical and Biological Research, H. Haseltine and N.L. First (Eds.), Liss, New York, pp. 67-68.

Bachvarova, R.F., Manova, K., Packer, A.I., Huang, E.J., and Besmer, P. (1993). In: Ovarian Cell Interactions. Genes to Physiology, A.J.W. Hsueh and D.W. Schomberg (Eds.), Serono Symposia, USA, Springer-Verlag, New York, pp. 25-37.

Bag Avant, H., Thillai-Koothan, P., Sharma, M.G., Talwar, G.P., and Gupta, S.K. (1994). J. Reprod. Fertil. 102, 17.

Baker, T.G. (1963). Proc. Roy. Soc. Ser. B., 158, 417.

Baker, T.G. and Franchi, L.L. (1967a). J. Cell Sci. 2, 213.

Baker, T.G. and Franchi, L.L. (1967b). Chromosome. 22, 358.

Baker, T.G. and Franchi, L.L. (1969). Z. Zellforsch. Mikrosk. Anat. 93, 45.

Baker, T.G. and O W.W. (1976). Clin. Obstet. Gynecol. 3, 1.

Bakker, J. and Baum, M.J. (2000). Front. Neuroendocrinol. 21, 22.

Balasch, J., Febregues, F., Creus, M., Penarrubla, J., Vidal, E., Carmona, F., Puerto, E., and Vanrell, J.A. (2000). J. Assist. Reprod. Genet. 17, 120.

Baldi, E., Luconi, M., Bonaccorsi, L., Muggi, M., Francavilla, S., Gabriele, A., Properzi, G., and Forti, G. (1999). Steroids 64, 143.

Baltsen, M. and Byskov, A.G. (1999). Biomed. Chromatogr. 13, 382.

Banos, M.E., Rosales, A.M., Ballesteros, L.M., Hernandez-Perez, O., and Rosado, A. (1996). Arch. Med. Res. 27, 49.

Baram, T.Z., Mitchell, W.G., Brunson, K., and Haden, E. (1999). Dev. Neurosci. 21, 281.

Barreca, A., Artini, P.G., Cesarone, A., Arvigo, M., D'Ambrogio, G., Genazzani, A.R., Glordano, O., and Minuto, F. (1996). J. Endocrinol. Invest. 19, 35.

Battaghia, D.E., Godwin, P., Klein, N.A., and Soules, M.R. (1996). Hum. Reprod. 11, 2217.

Bauskin, A.R., Franken, D.R., Eberspaecher, W., and Donner, P. (1999). Mol. Hum. Reprod. 5, 534.

Bavister, B., Cummins, J., and Poldan, R.S. (Eds.) (1990). Fertilization in Mammals. Serono Symposia, USA. Norwell.

Behrman, H.R., Preston, S.L., Pellicer, A., and Parmer, T.G. (1988). In: Meiotic Inhibition, Molecular Control of Meiosis, Progress in Clinical and Biological Research, H. Haseltine and N. First (Eds.), Lass, New York. pp. 115-136.

Belay, E.D., Bresee, J.S., Holman, R.C., Khan, A.S., Shahnari, A., and Schonberger, L.B. (1999). N. Engl. J. Med. 340, 1377.

Belin, D., Huarte, J., Vassalli, A., Sappino, P.A., Basco, D., Strickland, S., and Vassalli, J.D. (1988). In. Development and Function of the Reproductive Organs, M. Parvinen, I. Huhtaniemi and I.J. Pelliniemi, (Eds), Vol. 2, Ares-Serono Symposia, Rome, Italy. pp. 279-284.

Bennett, R.A., Osathanondh, R., and Jeh, J. (1996). J. Clin. Endocrinol. Metab. 81, 3073.

Benoit, A.M. and Veldhuis, J.D. (1995). Biol. Reprod. 53, 46.

Bergrant, A.M., Heim, K., Ulmer, H., and Illmensee, K. (1999). J. Psychosom. Res. 46, 391.

Berndtson, A.K., Vincent, S.E., and Fortune, J.E. (1995). J. Reprod. Fertil. 49, 527.

Bevers, M.M., Dieleman, S.J., Hurk, R., Vanden, Izadyar, F., and Van den Hurk, R. (1997). Theriogenelogy 47, 13.

Billig, H., Tornell, J., Carlsson, B., Hillensjo, T., and Ahren, K. (1988). In: Development and Function of the Reproductive Organs, M. Parvinen, I. Huhtaniemi, and L.J. Pelliniemi (Eds.), Vol. 2, Ares-Serono Symposia, Rome, Italy, pp. 161-170.

Blandau, R.J. (Ed.) (1975). Aging Gametes: Their Biology and Pathology. International Symposium on Aging Gametes, Seattle Wash. 1973, S. Karger, Basel.

Block, E. (1952). Acta Anatomica 14, 108.

Bonello, N., Mckie, K., Jasper, M., Andrew, L., Ross, N., Braybon, E., Braenntroen, M., and Norman, R. (1996). Biol. Reprod. 54, 435.

Born, G., and Mannaerta, B. (2000). Hum. Reprod. 15, 149.

Bowen, J.M., Dahl, G.E., Evans, N.P., Thurn, I.A., Wang, YueDong, Brown, M.B., Karsch, F.I., and Wang, Y.D. (1998). Endocrinol., Philadelphia, 139, 588.

Bowling, F.G., Harrison, E., Bryon, J., Cats, J., and McGill, J.J. (1999). Eur. J. Pediatr.158, 794.

Braude, P., Bolton, V., and Moore, S. (1988). Nature, London, 332, 459.

Briggs, D.A., Sharp, D.J., Miller, D., and Gosden, R.G. (1999). Mol. Hum. Reprod. 5, 1107.

Brown, S.E., Mandelin, E., Oehninger, S., Toner, J.P., Seppula, M., and Jones, H.W. (2000). Fertil. Steril. 74, 338.

Brun, N., Robitaille, Y., Grignon, A., Robinson, B.H., Mitchell, G.A., and Lambert, M. (1991). Am. J. Med. Genet. 84, 94.

Buccione, R., Schroeder, A.C., and Eppig, J.J. (1990). Biol. Reprod. 43, 543.

Bulletti, C., de Ziegler, D., Guller, S., and Levitz, M. (Eds.) (2001). Annals of the New York Academy of Sciences, New York, Vol. 943, pp. 1-376.

Byskov, A.G.S. (1969). Z. Zellforsck Mikrosk. Anat. 100, 285.

Byskov, A.G. and Hoyer, P.E. (1994). In: The Physiology of Reproduction, E. Knobil and J.D. Neill (Eds.), Raven Press, New York, pp. 487-540.

Byskov, A.G., Anderson, C.Y., Novdholm, L., Thogersen, H., Guohang, X., Wassarmann, O., and Anderson, J.V. (1995). Nature 374, 559.

Byskov, A.G., Anderson, C.Y., Hossanm, A., and Guohang, X. (1997). Mol. Reprod. Develop. 46, 296

Byskov, A.G., Andersen, C.V., Leonardsen, and Baltsen, M. (1999). J. Exp. Zool. 285, 237.

Cahill, D.J., and Hull, M.G. (2000). Hum. Reprod. Update 6, 56.

Calarco, P.G. (1995). Develop. Genet. 16, 36.

Can, A. and Albertini, D.F. (1997). Mol. Reprod. Dev. 46, 351.

Cantani, A. (1999). Eur. Rev. Med. Pharmacol. Sci. 3, 143.

Carolan, C., Lonergon, P., Monget, P., Monniaux, D., and Meamillod, P. (1996). Mol. Reprod. Dev. 43, 477.

Carroll, J., Jones, K.T., and Whitlingham, D.G. (1996). Rev. Reprod. 1, 137.

Cecconi, S., D Aurizio, R., and Colonna, R. (1996). J. Reprod. Fertil. 107, 207.

Chamberleu and Dean (1990).

Chabbert Buffet, N., Djakoure, C., Maitre, S.C., and Bouchard. P. (1998). Front. Neuroendocrinol. 19, 151.

Check, J.H., Check, M.L., and Katsoff, D. (2000a). Hum. Reprod. 15, 1709.

Check, J.H., Dietterich, C. Lurie, D., and Nazari, A. (2000b). Hum. Reprod. 15, 1679.

Chen, L.C. and Huang, J.L. (1999). TaiwanErh. KDI, Hsueh Hiu. Tsa. Chih 40, 341.

Chen, S.U., Chen, H.F., Lien Y.R., H.O., R.N., Chang, H.C., and Yang, Y.S. (2000a). Arch. Androl., 44, 197.

Chen, C.D., Chen, H.F., Lu, H.F., Chen, S.W., Ho, R.N., and Yang, Y.S. (2000b). Hum Reprod. 15, 1037.

Chian, R.C., Buckett, W.M., Too, L.L., and Tan, S.L. (1999). Fertil. Steril. 72, 639.

Chian, R.C., Buckett, W.M., Tullandi, T., and Tan, S.L. (2000). Hum. Reprod. 15, 165.

Christin-Maitre, S. and Bouchord, P. (1999). Ann. Endocrinol (Paris). 60, 118.

Christenson, L.K. and Stouffer, R.L. (1997). J. Clin. Endocrinol. Metabol. 82, 2135.

Cobo, A.C., Reguena, A., Neuspiller, P., Aragon, S.M., Mercoder, A., Navarro, J., Simon, C., Remohi, J., and Pellicer, A. (1999). Hum. Reprod. 14, 1864.

Cohen, J., Talansky, B.E., Grifo, J., and Maiter, H.E. (Eds.). (1992). Micromanipulation of Human Gametes and Embryos: Raven Press, New York.

Colgnie, D.C., and Murdock, W.J. (1997). Anim. Reprod. Sci. 47, 197.

Contencin, P. Gumpert, L. Sleiman, J. Possel, L. De, Gaudenar, I., and Adamsbaumi, C. (1999). Arch. Otolaryngol. Head Neck Surg. 125, 777.

Conway-Myert, B.A., and Steinkampf, M.P. (1996). Int. Apl. W0930496 Ai-3Oct 23pp.

Cook, C.L., Slow, Y., Taylor, S., and Fallot, M.R. (2000). Fertil. Steril. 73, 859.

Cooper, T.G. (1986). The Epididymis Sperm Maturation and Fertilization. Springer-Verlag, Berlin.

Coskun, S., Uzuman, M., Lin, Y.C., Frnidman, C.I., and Alak, B.M. (1995). Biol. Reprod. 53, 670.

Crain, J.L., Wiemer, K.E., Steuerwald, N., and Young, E.D. (1998). Am. J. Obstet. Gynecol. 179, 299.

Cross, N.L., Morales, P., Overstreet, J.W., and Hanson, F.W. (1988) Biol. Reprod. 38, 235.

Crozet, N. (1993). In: Reproduction in Mammals and Man (C. Thibault, M.C. Levasseur, and R.H.F. Hunter eds.) pp. 327-348 Ellipses, Paris.

Cummins, J.M., Breen, T.M., Harrison, K.L., Show, J.M., Wilson, L.M., and Hennessey. J.F. (1986). J. In Vitro Fertil. Embryo Transf. 3, 284.

Dain, L. Guerrero, H., Polok-Dlfried, E., and Tesone, M. (1995). Fertil. Steril. 64, 335.

Danforth, D.K. (1996). Am. J. Obstet. Gynecol. 172, 747.

Dean, J. (1993). In: Ovarian Cell Interactions, Genes to Physiology. A.J.W. Hsueh and D.W. Schomberg (Eds.). Serono Symposia, USA, Springer-Verlag, New York, pp. 49-59.

Dearborn. D.G., Yike, I., Sorenson, W.G., Miller, M.J., and Etzel, P.A. (1999). Environ. Health and Perspect. 107, Suppl. 3, 495.

De Laat, S.W., Bluemink, J.G., and Mummery, C.L. (eds). (1989). Cell to Cell Signals in Mammalian Development. NATOASI Series. Series H., Cell Biology, Vol. H26.

De Martino, C., Froncavilla, S., Fabbrini, A., and Accinni, L. (1989). In: Ultrastructure of Human Gametogenesis and Early Embryogenesis (V. Van Blerkom and P.M. Motta, eds.). pp. 1-32. Kluwer Academic Publishers, Boston.

De Placido, G., Alviggi, C., Mollo, A., Strina, I., Varricchio, M.T., and Molis, M. (2000). Hum. Reprod. 15, 17.

Deshmukh, U.S., Talwar, G.P., and Gupta, S.K. (1994). J. Clinic. Immunol. 14, 162.

Develioglu, O.H., Cox, B., Toner, J.P., Gehninger, S., and Muasher, S.J. (1999). Hum. Reprod. 14, 1168.

De Vos, A. and Van Steirteghem, A. (2000). Cells Tissues Organs 166, 220.

De Vos, A. Van de Velde, H., Joris, H., and Van Steirteyrteme, A. (1999). Hum. Reprod. 14, 1859.

Dhanju, C.K., Dhanoa, S.K., and Guraya, S.S. (1990). Ind. J. Exp. Biol. 28, 369.

Dhanju, C.K., Dhanoa, S.K., and Guraya, S.S. (1991). Indian J. Exp. Biol. 29, 721.

Dhanju, C.K., Dhanoa, S.K., and Guraya, S.S. (1992). Curr. Sci. 62, 958.

Dieldrich, K., Vander Ven, H., Al-Hasani, S., and Krebs, D. (1988). Hum. Reprod. 3, 39.

Dietl. J. (1989). The Mammalian Egg Coat. Structure and Function. Springer Verlag, Heidelberg, New York.

Ding, J., Rana, N., and Dmowski, W.P. (1999). Hum. Reprod. 140, 476.

Dissen, G.A., Hill, D.F., Costa, M.E., Dees, W.L., Lara, H.E., and Ojldu, S.R. (1996). Endocrinol 137, 196.

Downs, S.M. (1995). Dev. Biol. 167, 502.

Downs, S.S. (1996). In: The Ovary: Regulation, Dysfunction, Treatment, M. Filcon and C. Hamignt (Eds). Elsevier, Amsterdam, pp. 141-148.

Downs, S.M. (1997a). Biol. Reprod. 57, 54.

Downs, S.M. (1997b). MoL. Reprod. Dev. 46, 155.

Downs, S.M. Daniel, S.A.J., and Eppig, J.J. (1988). J. Exp. Zool. 245, 89.

Downs, S.M. and Mastropolo, A.M. (1997). Mol. Reprod 46, 551.

Driscoll, O.L., Tyler, J.P., Hangon, J.T., Fisher, P.R., Birdsall, M.A., and Knight, D.C. (2000). Hum. Reprod. 15, 1305.

Dunbar, B.S. and O' Rand, M.G. (Eds.). (1991). A Comparative Overview of Mammalian Fertilization, Plenum Press, New York.

Dvorak, M. and Tesarik, (1980). In: Ultrastructure of Human Ovary, P.M. Motta and E.S.E. Hafez (Eds). Martinus Nighoff Publishers, The Hague, pp. 121-137.

Edwards, R.G. (1970a). In: Holbrecht Proc. 6th World Congr. Fertil. Steril. p. 64.

Edwards, R.G. (1970b). Philos. Trans. Roy. Soc. B 259, 103.

Edwards, R.G. (1987). In: Future Aspects in Human *In Vitro* Fertilization (W. Feichtinger and P. Kemeter eds). pp. 244-250, 106 (1957). Springer-Verlag, Berlin.

Edwards, M.C., Johnson, J.L., Marriage, B., Graf, T.N., Coyne, K.E., Rajagopalan, K.V., and MacDonald (1999). Ophthalmology.

El-Budrawi, H. and Hafez, E.S.E. (1980). In: Biology of Ovary, P.M. Motta and E.S.E. Hafez (Eds.). Martinizhoff, The Netherlands, pp. 291-302.

El-Sheikh, M.M., Hussein, M., Sheikh, A.A., Fouad, S., El-Sheikh, R., and Al-Hasoni, S. (1999). Eur. J. Obstet. Gynecol. Reprod. Biol. 83, 81.

Elsenhauer, K.M., Chun, S.Y., Minami, S. Billig, H., and Hsueh, A.J.W. (1996). In: The Ovary: Regulation, Dysfunction and Treatment, M. Filcori and C.F. Lamigni (Eds.), Elsevier, Amsterdam. pp. 103-112.

Elvin, J.A., Yan, C., and Matzuk, M.M. (2000). Mol. Cell. Endocrinol. 159, 1.

Englert, Y., Van den Bergh, M. Delbaerce A., Devreker, F., Koring, I., Hannes, M., Emiliani, S., and Biramane, J., Vannin, A.S., Govaerts, I., Holoye, A. and Revelard, P. (1999). Rev. Med. Brux. 20, A 435.

Eppig, J.J. (1985). In: Developmental Biology. A Comprehensive Synthesis. Vol. l Sogenesis, I.W. Browder (Eds.). Plenum Press, New York, pp. 13-350.

Eppig, J.J. (1991). Biol. Reprod. 45, 824.

Eppig, J.J. (1994). Sem. Develop. Biol. 5, 51.

Eppig, J.J., O'Brien, M., and Wiggleswarth, K. (1996). Mol. Reprod. Develop. 44, 260.

Espane, F., Sanchez-Guenca, J., Fernandez, P.J., Gilbert, J., Romen, A., Estelles, A., Royo, M., and Muller, C.H. (1999). Andrologia 31, 1217.

Espey, L.L., and Lipner, H. (1994). In: The Physiology of Reproduction, E. Knobil, J.D. Neill et al. (Eds.). 2nd Edition, Vol I, Raven Press, New York pp. 725-780.

Fair, T., Hytell, P., Greve, T., and Boland, M. (1996). Mol. Reprod. Dev. 43, 50.

Fair, T., Hulshof, S.C.J., Hytell, P., Greve, T., and Boland, M. (1997). Mol. Reprod. Dev. 46, 208.

Fausor, B.C.J.M. (Editor-in-chief) and Rutherford A.J., Strauss, J.F. and Van Steirteghem (co-eds.) (1999). Molecular Biology in Medicine. The Parthenon Publishing Group, New York and London.

Feichtinger, W., and Kemetger, P. (Eds). (1987). Future Aspects of Human *in vitro* Fertilization. Springer-Verlag. Heidelberg.

Felberbaum, R.E., Albano, C., Ludwig, M., Riethmuller Winzen, M., Grigat, M., Devroly, P., and Driedrich, K. (2000). Hum. Reprod. 15, 1015.

Felix and Calhoun, D.A. (2000). Pediatrics 106, 340.

Feral, C., Bemhaien, A., and Leymaire, P. (1996). Europ. J. Endocrinol. 133: 747.

Ferrer-Canobate, J. Ros-Bullon, R., and Majado-Murtinez, J. (2000). Haematologica 65, 663.

Filnberg, R. and Cohen, R.B. (1965). Am. J. Obstet. Gynec. 92, 959.

Filicori, M. and Flamigni, C. (Eds). (1996). The Ovary: Regulation, Dysfunction and Treatment, Elsevier, Amsterdam.

Filicori, M., Cognigni, G.E., Taraborrelli, S., Spettoii, D., Clampaglia, W., and de-Fetis, C.T. (1999). Fertil. Steril. 72, 1118.

Findlay, J.K. (Ed.). Molecular Biology of the Female Reproductive System. Academic Press San Diego.

Findlay, J.K., Drummond, A.E. and Fry, R.C. (1996). Anim. Reprod. Sci. 42, 321.

First, N.L. and. Fields, M.J. (Eds.) (1994). In: New Advances in Reproductive Technologies p. 244.

Flemming, A.D., Khalil, W., and Armstrong, D. II (1983). J. Reprod. Fert. 69, 665.

Foote, R.H. (1975). In: The Aging Gametes, Their Biology and Pathology. R.G. Blanda (Ed.). International Symposium on Aging Gametes, Seattle, Wash. (1973). S. Karger, Basel, pp. 179-200.

Fried, G. and Wramsby, H. (1998). Hum. Reprod. 13, 656.

Frydman, P., Howles, C.M., and Truong, F. (2000). Hum. Reprod. 15, 520.

Fujiwara, T., Lambert-Messerlian, G., Sidis, Y., Leybin, L., Isaacson, K., Toth, T. and Schneyer, A. (2000). Fertil Steril. 74, 340.

Fukoyo, T. and Yajima, A. (1996). J. Clin. Endocrinol. Metab. 61, 828.

Fukui, Y., Fujii, K., Kumata, M., Imai, K., Alfonso, N.M., and Ono, H. (1986). Res. Bull. Obilnao Univ. Ser. 1 15, 27.

Fulop, C., Salustri, A., and Hascoll, V.C. (1997). Arch. Biochem. Biophy. 337, 261.

Funahashi, H., Contloy, T.C., and Day, B.N. (1997). Theriogenology 47, 679.

Fyfe, S., Leonard, H., Dye D., and Leonard, S. (1999). J. Child Neurol. 14, 440.

Galloway, S.M., Mc Natty, K.P., Cambridge, L.M., Laitinox, M.P., Jugengel, J.L., Jokiranta, T.S., Mc Laren, R.J., Luiro, K., Dodds, K.G., Montgomery, G.W., Beattie, A.E., Davis, G.H., and Ritvos, O. (2000). Nat. Genet. 25, 279.

Galway, A.B., OiKawa, M., Ny, T., and Hsueh, A.J. (1989). Endocrinology 125, 126.

Garrido, N. Navarro, J., Remohi, J., Simon, C., and Pellicer A. (2000), Hum. Reprod. Update. 4, 67.

Gentry, P.A., Plante, L., Sonrorder, M.O., La Marre, J., Young, J.E., and Dodds, W.G. (2000). Fertil. Steril. 73, 848.

Gilbert-Barness, E., and Barness, L.A. (1999). Pediatr. Dev. Pathol. 2, 286.

Gilchrist, R.B., Nayudn, P.L., Nowshari, M.A., and Hodges, J.K. (1995). Biol. Reprod. 52, 1234.

Gindoff, P.R., and Jewelewicz, R. (1986). Fertil. Steril. 46, 189.

Goetz, F.W., Berndtson, A.K., and Ranjon, M. (1991). In: Vertebrate Endocrinology: Fundamental and Biomedical Implications Vol. 4, Part A, Reproduction P.K.T. Pong and M.P. Schreibman (Eds.). Academic Press, San Diego, pp. 127-204.

Goden, R.G., and Fadaly, M.J. (1998). Reprod. Fertil. Dev. 10, 73.

Gomel, V., and Leung, P.C.K. (Eds.). 1997. In Vitro Fertilization and Assisted Reproduction. Proc. 10th World Congr. In Vitro Fertilization and Assist. Reprod., Vancouver (Canada) May 24-28. (1997).

Gondos, B., Bhiraleus, P., and Hobel, C.L. (1971). Am. J. Obstet. Gynecol. 110, 644.

Gondos, B., Westerguard, I., and Byskov, A.G. (1986). Am. J. Obstet. Gynecol. 155, 189.

Gook, D.A., Edgar, D.M., and Stern, C. (1999). Hum. Reprod. 14, 2061.

Gosden, R.G., and Bownes, M. (1995). In: Gametes-The Oocyte. J.G. Grudzinskas and J.L. Yovich (Eds.). Cambridge University Press, Cambridge. pp. 23-55.

Gosden, R., Oktay, K., Mullan, J., and Jenner, L. (1996). In: The Ovary: Regulation, Dysfunction and Treatment, M. Filicori and C. Flamigni (Eds.). Elsevier, Amsterdam, pp. 149-158.

Gossler, A. (1992). In: Early Embryonic Development of Animals W. Henning (ed). pp. 151-202. Springer-Verlag, Berlin.

Goud, P.F., Goud, A.P., Qlan, C. Leverge, H., Van der Elst J., De Sutter, P., and Dhont, M. (1998). Hum. Reprod. 13, 1638.

Goud, P., Goud, A., Van Oostveldt, P., Van der Elst, J., and Dhont, M. (1999a). Fertil. Steril. 72, 245.

Goud, P.T., Goud, A.P., Van Oostveldt, P., and Dhont, M. (1999b). Mol. Hum. Reprod. 5, 441.

Gougeon, A. (1994). In : The Ovary EY and C.C.K. Leung (Eds). Raven Press, New York, pp. 21-40.

Gougeon, A. (1996a). Endocr. Rev. 17, 121.

Gougeon, A. (1996b). In: The Ovary: Regulation, Dysfunction and Treatment (M. Fillicori and C. Flamingni (Eds). Elsevier, Amsterdam, pp. 3-12.

Gougeon, A. and Testart, J. (1986). J. Reprod. Fert. 78, 389.

Gregory, L. (1998). Hum. Reprod. 4, 117.

Grieshaber, N.A., Boitano, S., Ji, I., Mather, J.P., and Ji, T.H. (2000). Endocrinol. 141, 3461.

Grondohl, C., Hyttel, P., Grondohl, M.I., Eriksen, T., Gotfredsen, P., and Greve, T. (1995). Mol. Reprod. Develop. 42, 94.

Gull, I., Geva, E., Lerner, G., va, L., Lessing, J.B., Wolman, I., and Amit, A. (1999). Eur. J. Obstet. Gynecol. Reprod. Biol. 85, 225.

Gupta, S.K. (ed.) 1999. Reproductive Immunology, Narosa, New Delhi.

Gupta, S.K., Kaul, R., Rajalakshmi, S., Sahai, P., Yurewicz, E.C., and Sacco, A.G. (1994). J. Reprod. Immunol. 27, 241.

Gupta, S.K., Chadha, K., Harris, J.D., Yurewicz, E.G., Sacco, A.G., Kolluri, S.K., and Afzalpurkar, A. (1996). Biol. Reprod. 55, 41.

Gupta, S.K., Jethanandani, P., Afzalpurkar, A., Kaul, R., and Santhanam, R. (1997a). Hum. Reprod. Update 3, 311.

Gupta, S.K., Sharma, M., Behera, A.K., Bisht, R., and Kaul, R. (1997b). Biol. Reprod. 57, 532.

Gupta , S.K., Yurewicz, E.C., Sacco, A.G., Kaul, R., Jethanandani, P., and Govind, C.K. (1998). Mol. Hum. Reprod. 4, 1058.

Gupta. S.K., Shrivastava. N, Govind, C.K., Sivapuropu, N., and Gohlay, G.K. (2000). PINSA 66, 1.

Guraya, S.S. (1961). Quart. J. Microsc. Sci. 102, 381.

Guraya, S.S. (1966). Ani. J. Obstet. Gynecol. 96, 907.

Guraya, S.S. (1967a). Res. Bull. Punjab Univ. 18, 221.

Guraya, S.S. (1967b). Am. J. Obstet. Gynecol. 98, 99.

Guraya, S.S. (1967c). Nature (London) 714, 614.

Guraya, S.S. (1968). Am. J. Obstet. Gynecol. 101, 448.

Guraya, S.S. (1969). Acta Anat.74, 65.

Guraya, S.S. (1970a). Acta Anat. 77, 617.

Guraya, S.S. (1970b). Acta Embryol. Exp. 3, 227.

Guraya, S.S. (1970c). Acta Morphol. Neer. Scand. 7, 30.

Guraya, S.S. (1970d). Acta Morphol. Neer Scand. 7, 311.

Guraya, S.S. (1972a). Acta Eudocrinol. 69, 107.

Guraya, S.S. (1972b). Acta Anat. 81, 507.

Guraya, S.S. (1973). In: Proceedings of the Symposium. Ovogenesis-folliculogenesis. Nouzilly Ann. Biol. Anim. Biochem. Biophys. (Suppl) 13, 229.

Guraya, S.S. (1974). Int. Rev. Cytol. 37, 121.

Guraya, S.S. (1976a). Arch. Ital. Anat. Embriol. 81, 189.

Guraya, S.S. (1976b). Arch. Ital. Anat. Embriol. 81, 61.

Guraya, S.S. (1982). Int. Rev. Cytol. 78, 257.

Guraya, S.S. (1984). In: Human Fertility, Health and Nutrition: Impact of Molecular Biology and Biotechnology, J. David Puett (Ed.). Published by United Nations Funds for Population Activities, New York, pp. 103-128.

Guraya, S.S. (1985). Biology of Ovarian Follicles in Mammals. Springer-Verlag (Heidelberg, Berlin, New York).

Guraya, S.S. (1986). The cell and Molecular Biology of Fish Oogenesis. Karger, Basel.

Guraya, S.S. (1987). Biology of Spermatogenesis and Spermatozoa in Mammals. Springer-Verlag (Heidelberg, Berlin, New York, Tokyo).

Guraya, S.S. (1989). Ovarian Follicles in Reptiles and Birds. Zoophysiology, Vol.24. Springer-Verlag (Heidelberg, Berlin, New York).

Guraya, S.S. (1997a). Ovarian Biology in Buffaloes and Cattle. ICAR, New Delhi.

Guraya, S.S. (1997b). In: Frontiers in Environmental and Metabolic Endocrinology, S.K. Maitra (Ed.). The University of Burdwan, Burdwan, India, pp. 31-58.

Guraya, S.S. (1998a). Cellular and Molecular Biology of Gonadal Development and Maturation in Mammals: Fundamentals and Biolmedical Implications. Narosa Publishing House, New Delhi.

Guraya, S.S. (1998b). PINSA, B64, 161.

Guraya, S.S. (1999a). PINSA, B65, 163.

Guraya, S.S. (1999b). Comparative Testicular Biology in Animal Structure, Functions and Intercellular Communications. Science Publishers, New Hampshire, USA.

Guraya, S.S. (2000a). Comparative Cellular and Molecular Biology of Ovary in Mammals. Fundamentals and Applications. Science Publishers, New Hampshire, USA and Oxford IBH, New Delhi.

Guraya, S.S. (2000b). Inter. Rev. Cytol. 199, 1.

Guraya, S.S. and Greenwald, G.S. (1964). Anat. Rec. 149, 411.

Guraya, S.S. Kaur, P., and Sharma, R.K. (1991). Eur. J. Morphol. 29, 161.

Guraya, S.S. and Dhanju, C.K. (1992). Indian J. Exp. Biol. 30, 958.

Guraya, S.S. Stegner, H.E., and Pape, C. (1974). Cytobiol. 9, 100.

Gvakharia, M.D., Lipshnltz, L.I., and Lamb, D.J. (2000). Fertil. Steril. 73, 395.

Hamatani, T., Tanabe, K., Kamei, K., Sakai, N., Yamamoto, Y., and Yoshimura, Y. (2000). Biol. Reprod. 62, 1201.

Harada, M., Miyano, T., Matsumura, K., Osaki, S., Miyake, M., and Kato, S. (1997). Theriogenelogy 48, 743.

Harford, J.B. (Series Ed.) (1995). Modern Cell Biology Series. Holbrook: Cellular Ageing and Cell Death. John Wiley & Sons (Asia). Pvt. Ltd. Singapore.

Hassold, T., Abruzzo, M., Adkins, K., Griffin, D., Merrill M., Millie, E., Sakov, D., Shen, J., and Zaragoza, M. (1996). Mol. Mutagen, 28, 167.

Hazzouri, M., Rousseaux, S., Mongelard, F., Usson, Y., Pelletier, R., Faure, A.K., Vourch, C., and Sele B. (2000). Mol. Reprod. Dev. 55, 307.

He-Changh, Damiani, P., Parya, J.B., Fissore, N.A., and He, C.L. (1997). Biol. Reprod. 57, 1245.

Heikinheimo, O., and Gibbons, W.E. (1998). Mol. Hum. Reprod. 4, 745.

Hertig, A.T. (1968). Am. J. Anat. 122, 107.

Hertig, A.T., and Adams, E.C. (1967). J. Cell. Biol. 34, 647.

Himms-Hagen, J. (1999). Crit. Rev. Clin. Lab. Sci. 36, 575.

Hinrichs, K. and Williams, K.A. (1997). Biol. Reprod. 57, 377.

Hirashima, Y., Kobayashi, M., Gotoh, J., and Terao, T. (1997). Europ. J. Obstret. Gynecol. Reprod. Biol. 73, 79.

Hinsch, K.D., and Hinsch, E. (1999). Andrologia 31, 326.

Hintz, S.R., Vincent, J.A., Pitlick, P.T., Fineman, J.R., Steinhorn, R.H., Kin, G.E., and Benitz, W.E. (1999). J. Perintol. 19, 441.

Hirroi, M. (1996). Nippon Fiyinka Gakkai Zassin 48, 578.

Hirshfield, A. N. (1991). Int. Rev. Cytol. 124, 43.

Hirshfield, A. N. (Ed.) (1989). Growth Factors and the Ovary. Plenum Press, New York.

Hoek, K. H., Vander Woodhouse, C. M., Brannstrom, M., Norman, R. J., and van Der Hork, K. H. (1998). Biol. Reprod. 58, 1266.

Homburg, R. and Shelef, M. (1995). In : Gametes: The Oocyte. J. G. Grudzinskas and J. L. Yovich (Eds.). Cambridge University Press, Cambridge, pp. 277-291.

Hong, S. W., Chung, H. M., Lim, J. M., Yoon, T. K., Yee, B., and Cha, K. Y. (1999). pert. Steril. 72, 142.

Hovatta, O., Wright, C., Kaausz, T., Hardy, K., and Winston, R. M. (1999). Human Reprod. 14, 2519.

Howlett, S. K., and Bolton, V. N. (1985). J. Embryol. Exp. Morph. 87, 175.

Hsieh, Y. Y., Hsu, T. Y., Lee, C. C, Chang, C. C., Tsai, H. D., and Tsai, C. H. (1999). J. Reprod. Med. 44, 737.

Hsueh, A. J. W., Bicksak, T. A., Ha, X. C., Dahl, K. D., Fauser, B.C.J.M., Brenda Galway, A., Czekala, N., Pavlou, S. N., Papkoff, H., Keene, J., and Boime, J. (1989). Recent. Prog. Horm. Res. 45, 209.

Hsueh, A. J. W., Billig, H., and Tsafriri, A. (1994). Endocr. Rev. 15, 707.

Huang, F. J., Chang, S. V., Tsai, M. Y., Lin, Y. C., Kung, F. T., Wu, J. F., and Lu Y. J. (1999). J. Assist. Reprod. Genet. 16, 483.

Hue, I., Pedieu, T., Huneau, D., Ruffini, S., Gall, L., and Crozet, N. (1997). Mol. Reprod. Dev. 47, 222.

Hull, M.G.R. (1995). In : Gametes: The Oocyte, J. G. and J. L. Yovich (Eds.). Cambridge University Press, Cambridge pp. 291-325.

Hung, C. H., Cheng, S. N., Hua, Y. M., Wang, C. L. Chen, Y. H. and Yang, K. D. (1999). Taiwan Erh. Ko. I. Huck. Hui. Tsa. Chip 40, 128.

Hyttell, P., Xu, P. K., Smith, P. and Greve, T. (1986). J. Reprod. Fertil. 78, 615.

Ipsiroglu, O. S., Eichler, F., Stocklea-Ipsiroglu, S., and Trattmig, S. (1999). Pediatr. Neurol. 21, 739.

Jablonka Shariff, A. and Olson, L. M. (1988). Endocrinol., Philadelphia 139, 2944.

Jablonka Shariff, A., Reynolds, L. P., and Redmer, D. A. (1996). Biol. Reprod. 55, 693.

Jalabert, B., Fostier, A., Breton, B., and Weil, C. (1991). In: Vertebrate Endocrinology, Fundamentals and Biomedical Implications. P. K. T. Pang and M. P. Schreibman (Eds.). Vol 4, Part A Reproduction. Academic Press, San Diego, pp. 23-90.

Jennings, J. C., Moreland, K., and Peterson, C. M. (1996). Drugs 52, 313.

Jequier, A. M. (1995). In: Gametes: The Spermatozoan. J. G. Grudzinskus and J. L. Yovich (Eds.). Cambridge University Press, Cambridge, pp. 175-191.

Jethanandani, P., Santhanam, R., and Gupta, S. K. (1998). Mol. Reprod. Develop. 50, 223.

Ji, Y. Z., Bommel, M., Jonannet, P., and Wolf, J. P. (1997). Mol. Reprod. Develop. 47, 120.

Jia, X. C., Ny, T., and Heueh, A. J. (1990). Molec. Cell. Endocr. 68, 143.

Jones, R. E. (1984). Human Reproduction and Sexual Behaviour. Prentice-Hall INC., Englewood Cliffs. N. J.

Kably Ambe, A., Barren Vallejo, J., Gongora Rodriguez, A., Carballo Mondragon, E., and Anta Jaen, E. (1999). Gynecol. Obstetr. 67, 4081 in Spanish.

Kanayama, T., Sato, E., Shimoda, K., and Miyamoto, H. (1990). J. Jap. Surg. 52, 199.

Kanitz, W., Torner, H., Becker, F., Alm, H., and Kurth, J. (1997). Arch-fur Tierz 40. Suppl., 128.

Karakji, E. G., and Tsang, B. K. (1995). Mol. Cell. Endocrinol. 107, 105.

Karam, H., Vadenaire, O., Belour, M. F., Prigent, Sassy, C., Rakotosaiama, A., Clozel, M., Itakovitz, J., and Bruneval, P. (1999). Cell Tissue Res. 295, 101.

Kasson, B.G., Meiden, R., Davoren, J.B., and Hsueh, A.J.W. (1985). Endocrinology, 117, 1027.

Kaul, R., Afzalpurkar, A., and Gupta, S.K. (1997). Mol. Reprod. Develop. 47, 140.

Kaur, C., and Guraya, S.S. (1986). Curr. Sci., 55, 995.

Kaur, C., and Guraya, S.S. (1989). Indian J. Anim. Sci. 59, 1242.

Keefe, D.L., Niven-Faurchild, P.S., and Buradagunta, S. (1995). Fertil. Steril. 64, 577.

Kentrup, H., Altmuller, J., Pfaffle, R., and Heimann, G. (1999). Eur. J. Endocrinol. 141, 379.

Kerjeen, A., Poirot, C., Espelbein, S., and Jouennet, P. (1999). Hum. Reprod. 14, 1578.

Kessel, R.G. (1985). In: Oogenesis, L.W. Browder (Ed.). Plenum Press, New York, pp. 179-233.

Khorram, O., Shapiro, S.S., and Jones, J.M. (2000). Fertil. Steril. 74, 163.

Kim, A.H., Khanna, A., Aten, R.F., Raymond, F., Olive, D.T.L., and Behrman, H.R. (1996). Mol. Hum. Reprod. 2, 549.

Kim, N.H., and Menno, A.R. (Jr) (1995). Mol. Reprod. Dev. 40, 304.

Kim, C.H., Cho, Y.K., and Mok, J.E. (1996). Hum. Reprod. 11, 398.

King, P.R., and Coetzer, W.A. (1996). J. South Afr. Vet. Assoc. 57, 42.

King, P.R., and Coetzer, W.A. (1997). J. South Afr. Vet. Assoc. 68, 16.

King, W.A., Bousquet, D., Greve, T., and Goff, A.K. (1986). Acta Vet. Scand. 27, 267.

Kito, S. and Bavistez, B.D. (1997). Biol. Reprod. 56, 1281.

Kobayashi, R. (2000). Psychiatry Clin. Neurosci. 34, 235.

Koch, R.K. (1999). Am. Fam. Physician, 60, 1462.

Korkmazo, A., Yigit, S., Firat, M., and Oran, O. (1999). Pediatr. Radiol. 30, 323.

Kovacic, B., and Vaisavgevic, V. (2000). Mol. Reprod. Dev. 55, 197.

Kruip, T. A. M., Cran, D. G., Van Beneden, T. H., and Dieleman, S. J. (1983). Gamete Res. 8, 29.

Kruip, T. A., Bevers, M. M., and Kemp, B. (2000). Theriogenology 15, 611.

Kuleshova, L., Gianavoli, L., Megli, C., Fesserethi, A., and Trounson, A. (1999). Hum. Reprod. 14, 3077.

Kurilo, L. F. (1981). Hum. Genetics 17, 86.

Kupker, W., Diedrich, K., and Edwards, R. G. (1998). Hum. Reprod. 13, 20.

Kuzan, B. B., and Quinn, P. (1988). In: *In Vitro* Fertilization and Embryo Transfer: A Manual of Basic Techniques (D. P. Wolf, ed: B. D. Bavister, M. Gerrity and G. S. Kopf, assoc. eds.) pp. 301-347. Plenum Press, New York.

Kuzi, N., Sueoka, K., Miyazaki, T., Tanaka, M., Oda, T., Kobayashi, T., and Yoshimura, Y. (1996). Biol. Reprod. 55, 984.

Lachaux, A., Bouvier, R., Loras-Duclaux, I., Chappuis, J. P., Meneguzzi, G., and Ortonne, J. P. (1999). J. Pediatr. Gastroenterol. Nutr. 29, 393.

Lanzavechhia, G., and Mangioni, C. (1964). J. Micr. 3, 447.

Larsen, W. J., Wert, S. E., and Brunner, G. D. (1987). Dev. Biol. 122, 61.

Lau, C. P., Ledger, W. L., Groome, N. P., Barlou, D. H., and Muttukrishna, S. (1999). Hum. Reprod. 14, 2525.

Leibfried-Rutledge, L., and First, N.I. (1979). J. Anim. Sci. 48, 76.

Leibfried-Rutledge, M. L., Critser, E. S., Eyestone, W. H., Northey, D. L., and First, H. L. (1987). Biol. Reprod. 36, 376.

Leibfried-Rutledge, M. L., Florman, H. M., and First, N. L. (1989). In: Molecular Biology of Fertilization. H. Schatten and G. Schatten (Eds.). Academic Press, New York, pp. 259-301.

Lemyre, E., Russo, P., Melancon, S. B., Gugne, R., Potier, M., and Lambert, M. (1999). Am. J. Med. Genet. 82, 385.

Lenton, E., Soltan, A., Hewitt, J., Thomson, A., Davies, W., Ashraf, N., Sharma, V., Jenner, L., Ledger, W., and Mc Veigh, E. (2000). Hum. Reprod. 15, 1021.

Letterie, G. S., Coddington, C. C., Collins, R. L., and Merriam, G. R. (1996). Hum. Reprod. 11, 19.

Lewit, N. and Kol, S. (2000). Fertil. Steril. 74. 401.

Li, H., Gyllensten, H. B., Cui, X., Saiki, R. K., Erlich, H. A., and Arnhein, N. (1988). Nature 335, 414.

Liakos, A. M., Bradley, N. K., Magram, G., and Muszyxski, C. (2000). Neurol. Res. 22, 69.

Liang, L.F. and Dean, J. (1993). Development. Biology, 156, 399.

Liu, H. W., Lin, Y.C., Cha, C.F, Chang, S. Y., and Sun, G. H. (2000). Mol. Hum. Reprod. 6, 422.

Liu, Jian Min., Sirots, J., and Liu, J.M. (1998). Biol. Reprod. 58, 1527.

Liu, Yi Xin, Peng, X. R., Liu, H. Z., Cheng, Yij, Ny, T., Liu, Y. X., Peng, X. R., Liu, H. Z., Cheng, Y. J., and Ny. T. (1998). Biol. Reprod. 59, 409.

Ludwig, M., Strik, D., Al Hasani, S., and Diedrich, K. (1999). Eur. J. Obstet. Gynecol. Reprod. Biol. 87, 13.

Lussier, J.G. and Carruthers, T.D. (1987). Theriogenology 27, 253 (Abstract).

Magerkurth, C., Topfer-Petersen, E., Schwartz, P., and Michelmann, H. W. (1999). Hum. Reprod. 14, 1057.

Magottin, D.A. and Erickson, G.F. (1994). In: Molecular Biology of Female Reproductive System. J. K. Findloy (ed.). Academic Press, San Diego, pp. 39-61.

Mailhes, J.B. (1995). Mutation Res. 339, 155.

Majumdar, A.C., Katiyar, P.K., Singh, R., Taneja, V. K., and Bhat, P. N. (1988). Proc. II World Buffalo Congress, New Delhi, India, Dec. 12-18, 1988 Vol. I. p. 54. (Abs.).

Makabe, S., Nattola, S.A., and Motto, P.M. (1989). In : Ultrastructure of Human Gametogenesis and Early Embryogenesis. J. van Blerkom and P. M. Motta (Eds.). Kluwer Academic Publishers, Dordrecht, pp. 33-60.

Makabe, S., Nagur, T., Nattola, S. A., Boreda, J., and Motto, P. M. (1991). In : Ultrastructure of Ovary. S. Makabe and P. M. Motta (Eds.), Kluwer Academic Publishers, Dordrecht, pp. 1-28.

Malter, H.E. and Cohen, J. (1989). Fertil. Steril. 51, 139.

Mao, C. and Grimes, D.A. (1988). Am. J. Obstet. Gynecol. 159, 279.

Mattioli, M. (1992). In: Embryonic Development and Manipulation in Animal Production. A Lauria and F. Gondolfi (Eds.). Portland Press, London, pp. 17-28.

Mayatepek, E., Lindner, M., Zelezny, R., Lindner, W., Brandstetter, G. and Hoffmann, G. F. (1999). Neuropediatrics 30, 5.

Menezo et al. 1993.

Menezo, Y.R. and Ben-Khalifa, M. (1995). J. Assist. Reprod. and Genet., 12, 35.

Menezo Y.J.R. Veiga A., and Pouly, J.J. (2000). Theriogenology 53, 599.

Menezo, Y., and Borak, V. (2000). Human Reprod., Oxford, 15, 1776.

Meng, L., Rutledge, J., Zhu, Y., Kidder, G.M., Khamsi, F., and Armstrong, D.T. (1996). Mol. Reprod. Dev. 43, 228.

Mermillod, P., Onissaid, R., and Cognia, Y. (1999). J. Reprod. Fertil. Suppl. 54, 449.

Meraic, F., Mutsumoto, K., and Wolffe, A.P. (1997). J. Biol. Chem. 272, 1284.

Metzi, J. D., Elias, E. R., and Bernul, C.I. (1999). Pecing Clin. Electrophysiol. 22, 821.

Mlyano, T., Moor, R. M., Wooding, F B. P., and Shiroo, M. (1996). Mol. Reprod. Dev., 44, 408.

Mogas, T. Palamo, M.J., Izquierdo, M.D., and Paramio, M.T. (1997). Theriogenology 47, 1189.

Monk, M., Handyside, A., Hardy, K., and Whittingham, D. (1987). Lancet 2, 423.

Moor, R.M., and Gandolfi, F. (1987). J. Reprod. Fertil. Suppl. 34, 55.

Moor, R.M., Osborn, J.C., and Cresby, I.M. (1985). J. Reprod. Fertil. 74, 167.

Moor, R., Hutchings, A., Hawkins, C., and Jung, T. (1992). In: Embryonic Development and Manipulation in Animal Production. A. Lauria and F. Gondolfi (Eds.). Portland Press, London, pp. 1-15.

Moor, R. M., Dai, Y., Lee, C., and Fulka, J. Jr (1998). Human Reprod. update 4, 223.

Mori, T., Irahara, M., Saito, H., Ohno, Y., and Hosoi, E. (1985). Acta Endocr. 110, 408.

Motta, P.M., and Makabe, S. (1986). Eur. J. Obstet. Gynecol. Reprod. Biol. 22, 171.

Motta, P.M., Makabe, S., Naguro, T., and Currer, S. (1994). Arch. Histol. Cytol. 57, 369.

Motta, P.M., Nottola, S. A., Familiari, G., Macchikelli, G., Vizza, E., and Correr, S. (1995). Ital. J. Anat. Embryol. 100, 9.

Motta, P.M., Makabe, S., and Nottola, S.A. (1997a). Human Reprod. Update 3, 281.

Motta, P.M., Makabe, S., and Nottola, S.A. (1997b). J. Obstet. Gynecol. Reprod. Biol. 75, 5.

Moudgal, N.R., Shetty, G., Selvaraj, N., and Bhatnagar, A. S. (1996). J. Reprod. Fertil. Suppl. 50, 69.

Mullen, C. A., Thompson, J. N., Richard, L. A., and Chan, K. W. (2000). Bone Marrow Transplant. 25, 1093.

Murdoch, W. L., Peterson, T. A., Van Kirk, E. A., Vincent, D. L., and Insheep, E. K. (1986). Biol. Reprod. 35, 1187.

Murdoch, W. J., Colgin, D. C., and Ellis, J. A. (1997). J. Anim. Sci. 75, 1601.

Murilla, F.V. (1999). East Afr. Med. J. 76, 281.

Nagnta, Y., Honjon, K., Sonoda, M., Sumii, Y., Inove, Y., and Kawarbayashi, T. (1999). Fertil. Steril. 72, 235.

Nakamura, K., Fujiwara, H. Nakaayama, T., Goto, Y., Tochibana, T., Suginami, H., Hedo, M., Maedo, M., and Mori, T. (1996). Human Reprod. 11, 1952.

Netter, A. (1993). In: Reproduction in Mammals and Man (C. Thibault, M., C. Leva Seur and R. H. F. Hunter, eds.). pp. 627-642. Ellipses, Paris.

Neuber, E. and Powers, R.D. (2000). Human Reprod. 15, 171.

Newton, H., Pegg, D. E., Barrass, R., and Goden, R. G. (1999). J. Reprod. Fertil. 117, 27.

Nigg, B.A. (1997). Nature 386, 779.

Nilsson, B.O., Jin, M., Andersson, A.C., Sunderstrom, P., and Larsson, E. (1999). Virus Genes 18, 115.

Nishi, S., Hoshi, N., Kasahura, M., Ishibashi, T., and Fujimoti, S. (1999). Mol. Hum. Reprod. 5, 495.

Nussbaum, D.J. and Paather, R.S. (1995). Mol. Reprod. Develop. 41, 71.

Ocana-Quero, J.M., Moreno-Millan, M., Pinedo-Merlin, M., and Rodero-Franganillo, A. (1997). Arch de Zoot 46, 51.

Oilkawa, M. and Hsueh, A.J. (1989). Endocrinology 125, 2550.

Ohta, N., Saito, H., Kuzumaki, T., Takahashi, T., Ito, M.M., Saito, T., Nakahara, K., and Hiroi, M. (1999). Mol. Hum. Reprod. 5, 22.

Oktay, K., Newton, H., Aubard, Y., Salha, O., and Gosdex, R.G. (1998). Fertil. Steril. 69, 1.

Oosterhuis, G.J., Vermes, I., Lambalk, G.E., Michelsen, H.W., and Schoemaber, J. (1998). Hum. Reprod. 13, 283.

Orvieto, R., Ben-Rafael, Z., Abir, R., Bar-Hava, I., Fisch, B. and Moland, Y. (1999). Am. J. Reprod. Immunol. 42, 288.

Osterlund, C. and Fried, G. (2000). Mol. Hum. Reprod. 6, 498.

Osuga, Y., Teutseoumi, O., Momoleda, M., Okagaki, R., Matsumi, H., Suenaga, A., Yano, T., and Taketani, Y. (1999). Mol. Hum. Reprod. 5, 703.

Palasz, A.T., and Mapletoff, R.J. (1996). Rec. Advanc. Biotechnol. Adv 14, 127.

Paoloni-Giacobino, A., Dahoun, S.P., Sizonenko. P.C., Stalberg, A., Chardonnens, D. and Campana, A. (2000). Gynecol. Endocrinol. 14, 105.

Paulson, J.R., Hatch, I.B., Lobo, R.A., and Sauer, M. V. (1997). Hum. Reprod. (Oxford) 12, 835.

Payton, B.V., and Bachvarova, R. (1990). In: Fertilization in Mammals. B.D. Bavister, J. Cummins and E.R.S. Roldan (Eds.). Serono Symposia USA Noawell, pp. 25-34.

Pawshe, C.H., Appa Rao, K.B.C., and Totey, S.M. (1998). Mol. Reprod. Develop. 49, 277.

Pellicer, A., Gaitan, P., Neuspiller, F., Ardiles, G., Albert, C., Remohi, J., and Simon, C. (1998). J. Reprod, Immunol. 39, 29.

Penarrubia, J., Balasch, J., Fabregnes, F., Creus, M., Civico, S.A., and Vanrell, J. (1999). Hum. Reprod, 14, 1703.

Pereda, J. and Motta, P. (1991). Arch. Histol. Cytol. 54, 419.

Peterson, M., Jennings, Z.A., van-Dulin, M., and Aitken, R.J. (2000). Cells-Tissues Organs 166, 228.

Peterson, M., Wilson, M.R., Morris, K.D., Van Duinon and Aitken, R.J. (1998). Am. J. Reprod. Immunol. 40, 198.

Pickett, B.P. and Ahlstrom, K. (1999). Otolaryngol. Clin. North. Am. 32, 1019.

Piontek, C.M., Baab, S., Peindl, K.S. and Wisner, K.L. (2000). J. Clin. Psychiatry 61, 170.

Plachot, M. (1995). In: Gametes: The Oocyte J.G., Gadzinskas and J.L. Yuvich (Eds.). Cambridge University Press, Cambridge. pp. 95-105.

Plachot, M., Veiga, A., Montagut, J., de Gronchy, J., Calderon, G-Lepretre, S., Junca, A.M., Santalo, J., Carles, E., Mandelbaum, J., Barri, P., Degoy, Cohen, J., Egozone, J., Sabatier, J.C. and Salat-Baroux, J. (1988). Hum. Reprod. 3, 627.

Plachot, M., Grouchy, J. de Junca, A.M., Mandelbaum, J., Cohen, J., and Salat-Baroux, J. (1990). In: Gamete Physiology R.H. Ash, J.P. BaJmaceda and I. Johnston (Eds.). Serono Symposia, USA, Nerwelb, pp. 145-152.

Plachot, M. and Popesan, P. (1993). In: Reproduction in Mammals and Man (C. Thiabault, M.C. Levasseur and R.H.F. Hunter eds.). pp. 715-730. Ellipses, Paris.

Porcu, E., Fabbri, R., Petracchi, S., Clotti, P.M., and Flamigni, C. (1999). Am. J. Obstet. Gynecol. 180, 1044.

Qu, J., Nisolle, M. and Donnez, J. (2000a). Fertil. Steril. 74, 113.

Qu, J., Godin, P.A., Nisolle, M., and Donnez, J. (2000b). Hum. Reprod. 15, 302.

Rabinoviu, J. and Jaffe, R.B. (1990). Endocr. Reviews 11, 532.

Ragni, G., Pilom, S., Rossi, P., Caridelli, S., De Lauretis, L., Vegetti, W. and Crosignani, P.G. (1999). Gynecol. Obstet. Invest. 47,151.

Rajalakshmi, M., and Griffin, P.D. (Eds). (1999). Male Contraceptions Present and Future. New Age International Publishers, New Delhi.

Rance, G., Beer, D.E., Cone-Wason, B., Shepherd, R.K., Dowell, R.C., King, A.M., Rickarde, T.W., and Clark, G.M. (1999). Ear Hear 20, 238.

Rawe, V. Y., Olmedo, S.S., Nodar, F.N., Doncel, G.D., Acosta, A.A., and Vitullo, A.D. (2000). Mol. Hum. Reprod. 6, 510.

Reeka, N., Berg, F.D., and Brucker, C. (1998). Hum. Reprod. 13, 2199.

Renault, F., Verstichel, P., Ploussard, J.P., and Costil, J. (1999). Muscle Nerve, 22, 252.

Ralph, J.H., Telfer, E.E., and Wilmut, I. (1995). Mol. Reprod. Develop. 12, 248.

Richard, F.J. and Sirard, M.A. (1998). Molec. Reprod. Dev. 50, 200.

Robertson, L., Wolf, D.P., and Tash, J.S. (1988). Biol. Reprod. 39, 397.

Roman, E., Artoz, A., Smitz, J.E., Faguer, B., Camus, M., Van Steirteghem, A.C., and Devroey, P. (2000). Hum. Reprod. 15, 1435.

Rongieres-Bertrand, C., Olivennes, F., Righini, C., Fachin, R., Taieb, J., Hamamaih, S., Bouchard, P., and Frydman. R. (1999a). Hum. Reprod. 14, 603.

Rongieres-Bertrand, C., Olivennes, F., Righini, C., Fachin, R., Taieb, J., Hamamaih, S., Bouchard, P., and Frydman. R. (1999b). Hum. Reprod. 14, 683.

Rose-Hollekant, T.A. and Bavister, B.D. (1996). Mol. Reprod. Dev. 44, 241.

Roy, C.A. and Russell, R.C. (2000). J. Am. Acad. Child. Adolesc. Psychiatry 39, 207.

Runesson. E., Bostroem, E.K., Janson, P.O., and Bralmstroem, M. (1996). Mol. Hum. Reprod. 2, 245.

Russo, V.E.A., Brody, S., Cove, D., and Ottolenghi, S. (Eds.). (1992). Development. The Molecular Genetic Approach. Springer-Verlag, Berlin.

Rutherford, A.J. and Gosden, R.G. (1999). Acta Pediatr. Suppl. 88, 13.

Saenz, R.B. (1999). Am. Fam. Physician 59, 381.

Sagata, N. (1996). Trends Cell. Biol. 6, 22.

Sahin, M. and du Plessis, A.J. (1999). Pediatr. Neurol. 21, 674.

Salha, O., Abushei Kha, N., and Sharma, V. (1998). Hum. Reprod. Update 4, 816.

Salustri, A., Yanagishita, M., and Mascal, V.C. (1990). Dev. Biol. 138, 26.

Satav, J.G. and Nair, C.K.K. (1990). PINSA B 65, 13.

Schats, R., Sutter, P.D., Bassil, S., Kremer, J.A., Tournaye, H., and Donnez, J. (2000). Hum. Reprod. 15, 169.

Schramm, R.D. and Bavister, B.D. (1999). Hum. Reprod. 14, 2544.

Schultz, R.M. (1988). In Meiosis Inhibition, Molecular Control of Meiosis Progress in Clinical and Biological Research, H. Haseltine and N.L. First (Eds.). Liss, New York, pp. 137-152.

Schuetz, A. (1990). In: Mechanism of Fertilization : Plants to Humans, NATO/AST Series H: Cell Biology, B., Dale (Ed.) Vol. 45, Springer-Verlag, Heidelberg, pp. 45-64.

Sehuaman, R.D. and Bavister, B.D. (1999). Hum. Reprod. 14, 2544.

Semba, S., Moriya, T., Youssef, E.M., and Sasano, H. (2000). Pathol. Int. 50, 549.

Shany, E., Saada, A., Jendon, D., Shaag, A., Hershkovitz, E., and Elpeleg, O.N. (1999). Biochem. Biophys. Res. Commun. 262, 103.

Sharma, R.K. and Vats, R. (2003). Influence of Habitation and defects in human population of North India. Ass. Annual J. ARPAN 2, 52-59.

Sharma, R.K. (2000). Spinal Dysraphism. In Proceeding of Refresher Course. P.U. Chandigarh. pp. 71-74.

Sharma, R.K., Guraya, S.S., and M.B., Sharma (2000). Biology of ovarian follicles in the goat: A review Ind: J. Anim. Sci. 70. (4), 369-385.

Sharma, R.K. and Guraya, S.S. (1990). Granulosa heterogenicity: A histochemical, lectin staining and scanning electron microscopic studies in Rattus rattus ovary. Acta Embr. Morphol Exp., 11(2): 107-129.

Sharma, R.K. and Guraya, S.S. (1995). Zona Pellucida : A histochemical lectin staining and scanning electron microscopic study during folliculogenesis in house rat (Rattus rattus). J. Anim. Morphol. Physiol., 42(182): 33-38.

Sharma, R.K. and Chowdhury, S. (1998). Ultrastructure changes in oocytes of caprine antral follicles. Indian J. Anim. Sci., 68(4): 332-336.

Sharma, R.K. and Sawhney, A.K. (1999). Fine morphology of membrana granulosa in caprine ovary. Indian J. Anim. Sci. 69: 109-113.

Sharma, R.K. (2003) Structural analysis of cumulus and corona cells of goat antral follicles: Possible functional significance. Indian Journal of Animal sciences 73(1): 28-32, January 2003.

Shimada, T., Morita, T., Nogai, K., Sato, F., and Campbell, G.R. (1993). In: Aging of Reproductive Organs. Y. Taketani and S. Kawa goe, (Eds.). Hum. Res. (Suppl. I). pp. 9-15.

Shinetuge, B., Runesson, E., Bonello, N.P., Brannstrom, M., and Norman, R.J. (1999). Hum. Reprod. 14, 302.

Shimamoto, T., Yamoto, M., and Nakano, R. (1998). Eu. J. Endocrinol. 138, 594.

Shim-Chan Seob, Lee-Sang, G.U., Song-Wookeum, Lee-Chulsang, Lu-Kyung Kwang, Shim, C.S., Lee, S.G., Song, W.K., Lee, C.S., Lee, K.K., and Kim, K.J. (1997). Mol. Reprod. Dev. 48, 185.

Sidhu, K. S. and Guraya, S.S. (1991). Int. Rev. Cytol. 127, 253.

Sills, E.S., Moomjy, M., Zaningvic, N., Veeck, L.L., Mc Gee, M., Palermo, G.D., and Rosenwaks, Z. (2000a). Hum. Reprod 15, 891.

Sills, E.S., Perloe, M., and Palermo, G.D. (2000b). Eur. J. Obstet. Gynecol. Reprod. Biol. 91, 133.

Singh, S.K., Agarwal, S.K., Yadav, M.C., and Shankar, U. (1998). Buffalo J. 14, 53.

Sirois, J. and Dore, M. (1997). Endocrinol. Philad 138, 4427.

Sirotkin, A.V., Teradajnik, T.E., Makarwich, A.V., and Bulla, J. (1998). Anim. Reprod. Sci. 51, 333.

Smith, M. and Sperling, D. (1999). Am. J. Med. Genet. 84, 346.

Smith, S.D., Mikkelsen, A., and Lindenberg, S. (2000). Fertil. Steril. 73, 541.

Smitz, J., Cortvrindt, R., and Hu, Y. (1998). Hum, Reprod. 13, 664.

Snell, W. and White, J.M. (1996). Cell (Cambridge, Muss) 85, 623.

Sobetzko, D., Eich, G., Kalff-Suska, M., Grezeschik, K.H., and Superti-Furga, A. (2000). Am. J. Med. Genet. 31, 239.

Srivastava, T., Garola, R.E., and Mellerstein, S. (1999). Pediatr. Nephrol. 13, 481.

Staessen, C., Camus, M., Clasen, K., De Vos, A., and Steirteghem, A. (1999). Hum. Reprod. 14, 2474.

Stegner, H.E. (1967). Anatomie und Entwicklugs geschichte 39, 1.

Stegner, H.E. and Wartenberg, H.H. (1968). Arch. Gynec. 199, 151.

Stein, A., Woolley, H., and Mc Pherson, K. (1999). Br. J. Psychiatry. 175, 455.

Stevenson, A.F. (1999). Indian J. Exp. Biol. 37, 1167.

Stock, A.E., Woodruff, T.K., and Smith, L.C. (1997). Biol. Reprod. 56, 1559.

Suganama, N., Kitagawa, T., Nawa, A., and Tomoda, V. (1993). In: Aging of Reproductive Organs. Y. Taketani and S. Kawagoe (Eds.). Horm. Res. 39 (Suppl 1), 16-21.

Suikkari, A.M., Tulppala, M., Tuuri, J., Hovatta, O., and Barnes, F. (2000). Hum. Reprod. 15, 747.

Sutovsky, P. and Schatten, G. (1997). Biol. Reprod. 56, 1503.

Suzuki, K. (1993). In: Aging of Reproductive Organs (Y. Takettani and S. Kawagoe, eds.). Horm. Res. (Suppl. I) 5-8.

Szell, A.Z., Kausche, A., May, K., Lacham, Kuplan, O., Lotatgis, N., Leeton, J.E., and Trounson, A. (1996). In: The Ovary: Regulation, Dysfunction and Treatment, M. Filicori and C. Flamigni (eds.). Elsevier, Amsterdam, pp. 159-168.

Taieb, F, Thiber, C., and Jessus, C. (1997). Mol. Reprod. Dev. 48, 125.

Tain, C.F., Goh, V.H., and Ny, S.C. (2000). Mol. Reprod. Dev. 55, 104.

Takahashi, M., Koids, S.S., and Donahol, P.K. (1986a). Molec. Cell. Endocr. 47, 225.

Takahashi, M., Mayashi, M., Manganaro, T.F., and Donahoe, T.K. (1986). Biol. Reprod. 35, 447.

Takahashi, I., Takahashi, T., Shoji, Y., and Takeda, G. (2000). Clin. Endocrinol. 53, 127.

Taketami, Y. and Kawagoe, S. (Eds.) (1993). Aging of Reproductive Organs. Horm. Res. (Suppl I). Karger, Basel.

Tanguay, P.E. (2000). J. Am. Acad. Child. Adolesc. Psychiatry 39, 1079.

Tarin, J.J. (1996). Hum. Reprod. 10, 1563.

Tesarik, J. (1987). In: Future Aspects in Human In Vitro Fertilization (W. Feichtinger and P. Kemeter eds.). pp. 251-261, Springer-Verlag, Berlin.

Tesarik, J. (1999). J. Assist. Reprod. Genet. 16, 216.

Tesarik, J. and Mondoza, C. (1999). Bioassays 21, 791.

Tevelde, E.R., Pearson, P.L., and Broekma, F.J. (Eds.) (2000). Studies in Profertility Series Volume 9, Female Reproductive Aging, The Parthenon Publishing Group, New York.

Thibault, C. and Levasseur, M.C. (1988). Hum. Reprod. 3, 513.

Thibault, C., Gerard, M., and Menezo, Y. (1976). Prog. Reprod. Biol. 1, 233.

Thibault, C., Szollosi, D., and Gerard, M. (1987). Reprod. Nutr. Develop. 27, 865.

Thibault, C., Levasseur, M.C., and Hunter, R.H.F. (Eds.). (1993). Reproduction in Mammals and Man. Ellipses, Paris.

Tirone, E., D'Alessandris, C., Hascall, V.C., Siracusa, G., and Salustri, A. (1997). J. Biol. Chem. 272, 4787.

Tok, J., Garzon, M.C., Cserhalmi-Friedman, P., Lam, H.M., Spitz, J.L., and Christiano, A.M. (1999). Exp. Dermatol. 8, 128.

Totan, M. and Albayrak, D. (1999). Acta Paediatr. 88, 342.

Toth-Heyn, P., Drukker, A., and Guignard, J.F. (2000). Pediatr. Nephrol. 14, 227.

Totey, S. M., Singh, G., Taneja, M.O., Pawshe, C.H., and Talwar, G.P. (1992). J. Reprod. Fert. 95, 597.

Treacy, E.P., Lambert, D.M., Barnes, R., Boriack, R.L., Vockely, J., O'Brien, L.K., Jones, P.M., and Bennett, M.J. (2000). J. Pediatr. 137, 1257.

Trounson, A. (1985). In: Developmental Biology. A Comprehensive Synthesis. Vol. 4, Manipulation of Mammalian Development, R.B.L. Gwattkin (Ed.). Plenum Press, New York pp. 149-194.

Trounson, A. and Bongsu, A. (1996). Curr. Top. Des. Biol. 32, 59.

Tsafriri, A. and Adashi, E.Y. (1994). In: The Physiology of Reproduction, E. Knobil, J.D. Neill et al. (Eds.). 2nd Edition Vol. I, Raven Press, New York pp. 817-860.

Tsafriri, A., Reichh, R., and Abisogun, A.O. (1987). In: Marshall's Physiology of Reproduction, G.E. Lemming (Ed.). Vol. 3, Churchill Livingstone, London.

Tsai, M.C., Takeuchi, T., Bedford, J.M., Reis, M.M., Rosenwaks, Z., and Pabermo, G.D. (2000). Hum. Reprod. 15, 988.

Ujioka, T., Matsukawa, A., Tanaka, N., Matsura, K., Yoshinaga, M., and Okamura, H. (1998). Biol. Reprod. 58, 526.

Vats, R. (1999). Ecogenetic Investigations on Neural Tube Defects of North India Population Ph.D. Thesis submitted to Kurukshetra University.

Yen, S.S.C., Jaffe, R.B., and Barbieri, Kurukshetra, R.L. (Eds.) (1999). W.B. Saunders Company, Philadelphia.

Wolf, D.P. and Zelinoobi, Wooten, M. (Eds.) (2001). Assisted Fertilization and Nuclear Transfer in Mammals, Human Press, Totowa, New Jersey.

Yoshimura, Y., Karube, M., Aoki, H., Aoki, H., Oda, T., Koyama, N., Nagai, A., Akindo, Y., Hirano, H., and Nakamura, Y. (1996a). Endocrinology 137, 1204.

Yoshimura, Y., Ando, M., Nagamatsu, S., Iwashita, M., Adachi, T., Sulka, K., Miyazaki, T., Kuji, N., and Tanaka, M. (1996b). Biol. Reprod. 55, 152.

Yoshimura, Y., Nagamatsu, S., Ando, M., Washita, M., Oda, T., Katsumata, Y., Shiokawa, S., and Nakamura, Y. (1996c). Endocrinology 137, 438.

Zamboni, L. (1987). Fertil. Steril. 48, 711.

Zamboni, L., Mishell, D.R., Jr., Bell, J.H., and Baca, M. (1966). J. Cell. Biol. 30, 579.

Zanagnolo, V., Dharmarajan, A.M., Erdo, K. and Wallach, E.E. (1996). Fertil. Steril. 65, 1036.

Zenzes, M.T. (2000). Hum. Reprod. Update 6, 122.

Zernicka-Goetz, M., Verlhoc, M.H., Geraud, G., and Kubiock, J.Z. (1997). Europ. J. Cell. Biol. 72, 30.

Zhang, J., Wang, C.W., Krey, L., Liu, H., Neng, L., Elaszevyk, A., Adler, A., and Grifo, J. (1999). Fertil. Steril. 71, 729.

Zhao, J., Dorland, M., Taverne, M.A., Van der Weh-Jden, G.C., Bevers, M.M., Van Den-Hurk, R. (2000). Mol. Reprod. Dev. 55, 65.

Zullo, F., Pellicano, M., Zupi, E., Mastaantonio, P., and Happi, C. (2000). Fertil. Steril. 74, 376.

Subject Index

F

G

H

I

K

L

M

N